*Biochemical
Measurement and
Control*

Instrument
Society of
America

Gregory K. McMillan

Biochemical Measurement and Control

INSTRUMENT SOCIETY OF AMERICA

Biochemical Measurement and Control

© by the Instrument Society of America 1987
All rights reserved.
Printed in the United States of America.
No part of this publication may be reproduced,
stored in a retrieval system,
or transmitted, in any form or by any means,
electronic, mechanical, photocopying, recording or otherwise,
without the prior written permission of the publisher:

    The Instrument Society of America
    67 Alexander Drive
    P.O. Box 12277
    Research Triangle Park, NC 27709

Library of Congress Cataloging in Publication Data

McMillan, Gregory K., 1946 –
    Biochemical measurement and control.
    Bibliography: p.
    1. Biochemical engineering — technique.
    2. Chemical process control.
I. Title.
TP248.3.M36 1987   660'.63   87-2954
ISBN 0-87664-942-8

Contents

Figures   vii

Tables   x

Foreword   xi

Acknowledgments   xiii

Chapter 1 — An Industrial Perspective   1
    1.1   Products   2
    1.2   Process Characteristics   8
    1.3   Measurement and Control Characteristics   18

Chapter 2 — Flow, Level, Pressure, and Temperature Measurements   31
    2.1   Flow Measurements   32
    2.2   Level Measurements   35
    2.3   Pressure Measurements   36
    2.4   Temperature Measurements   36

Chapter 3 — REDOX and pH Measurements		**39**
3.1	REDOX Measurements	39
3.2	pH Measurement	44
3.3	Electrode Installation	60
3.4	Electrode Sterilization	62
Chapter 4 — Dissolved Oxygen and Carbon Dioxide Measurements		**65**
4.1	Dissolved Oxygen Measurement	66
4.2	Dissolved Carbon Dioxide Measurement	71
Chapter 5 — Composition Measurements		**77**
5.1	Biosensors	78
5.2	CHEMFETs and ISFETs	81
5.3	Fiber Optic Probes	83
5.4	Light Scattering	86
5.5	Nuclear Magnetic Resonance	88
5.6	Silicone Tubing Probes	90
5.7	Mass Spectrometers	91
Chapter 6 — Control Systems		**92**
6.1	Error Prediction	102
6.2	Effect of Process Equipment Type	107
6.3	Critical Loop Characteristics	108
Chapter 7 — Computer Systems		**123**
7.1	FDA Audits	124
7.2	Automated Experimentation	125
7.3	Inferred Measurements	125
7.4	Optimization	130
References		**133**
Appendix A — Continuous Biochemical Reactor Composition Control Example		**137**
Appendix B — ACSL Listing for Dynamic Simulation of Fermentor		**141**

Figures

1.1	The Living Cell as a Catalytic Reactor	1
1.2	Chemical Intermediates from Biomass by Gasification	7
1.3	Recombinant DNA (Genetic Engineering)	9
1.4	Feed Preparation and Reaction	13
1.5	Batch Cell Growth and Product Formation Curves	14
1.6	Product Separation and Purification (Cellular, Intracellular, and Extracellular Products	16
1.7	Peak Control Error for a Load Disturbance	21
1.8	Stirred Fermentor Measurement and Control	25
1.9	Loop Fermentor Measurement and Control	26
2.1	Vortex Meter Coefficient	34
2.2	Dynamic Temperature Error	37
3.1(a)	Generic REDOX Curves	41
3.1(b)	Generic pH Curve	41
3.2	REDOX Potential of Pt/Pt0 System	44
3.3	Equivalent Electrical Circuit for pH Measurement System	48
3.4	Horizontal Shift of Isopotential Point	49

3.5	Vertical Shift of Isopotential Point	*51*
3.6	Measurement Electrode Breakage Results in a Constant Millivoltage Corresponding to 5 to 6 pH Reading	*57*
3.7	Insertion Electrode Assembly for Biochemical Applications	*61*
4.1	Typical Current Voltage Curves for DO Electrode	*67*
4.2	Construction of Typical Dissolved Oxygen Electrode	*70*
4.3	Effect of pH on Solubility of Carbon Dioxide	*74*
4.4	Construction of an Industrial Dissolved Carbon Dioxide Probe	*75*
5.1	On-line System for Glucose Analysis	*80*
5.2	CHEMFET or ISFET Configuration	*82*
5.3	Fiber Optic Probe Construction	*83*
5.4	Fiber Optic Probe Output	*84*
5.5	Cell Culture Fluorescence	*86*
5.6(a)	Optical Density and Cell Dry Weight	*87*
5.6(b)	Optical Density and Cell Dry Weight with Protein Accumulation	*87*
5.7	In-line Nuclear Magnetic Resonance Sensor	*90*
5.8	Silicone Tubing Probe for Fermentor	*91*
5.9	Internal Configuration of a Mass Spectrometer	*92*
6.1	Packless Sanitary Control Valve	*98*
6.2	Block Diagram of a Generic Control Loop	*101*
6.3	The Effect of Equipment Type on Composition Control	*108*
6.4	Fed Batch 1500 Liters Stirred Fermentor/ Fermentor pH/Temperature	*110*
6.5	Fed Batch 1500 Liters Stirred Fermentor/ Dissolved Oxygen/Agitator Speed	*111*
6.6	Fed Batch 1500 Liters Stirred Fermentor/ Oxygen in Off-gas/Carbon Dioxide in Off-gas	*112*
6.7	Fed Batch 1500 Liters Stirred Fermentor/ Oxygen Uptake Rate/Carbon Dioxide Prod Rate	*113*
6.8	Fed Batch 1500 Liters Stirred Fermentor/ Gas Flow/Substrate Feed	*114*
6.9	Fed Batch 1500 Liters Stirred Fermentor/ Cell Conc/Growth Rate	*115*

6.10	Fed Batch 1500 Liters Stirred Fermentor/ Dilution Flow/Feed Conc	*116*
6.11	Fed Batch 1500 Liters Loop Fermentor/ Fermentor pH/Temperature	*117*
6.12	Fed Batch 1500 Liters Loop Fermentor/ Fermentor pH/Temperature/ Fast pH, DO, and Temperature Sensors	*118*
7.1	Laboratory Example Results for Inferred Measurements	*126*
7.2	Specific Growth Rate as Function of Substrate Concentration	*131*
7.3	Yield Maximization for Continuous Fermentor	*131*

Tables

1.1	Projected Annual Bulk Sales of Biotechnology Products by the Year 2000	3
1.2	New Biochemical Drugs Being Tested	5
1.3	Glucose and Oxygen Rangeability Requirements for E coli	15
1.4	Relative Magnitudes of Errors in Biochemical Processes	22
1.5	Typical Stirred and Loop Fermentor Control Loops	28
2.1	Typical Flow Measurement Methods	33
2.2	Typical Time Constants for Thermocouples in Thermowells	38
2.3	Typical Time Constants for Bare Temperature Elements	38
3.1	pH Measurement Errors Part 1	52
	Part 2	53
3.2	Liquid Junction Potentials	55
3.3	Sterilization-induced REDOX and pH Measurement Problems	63
4.1	Flow Sensitivity of DO Electrodes	69
5.1	Typical Response Characteristics of Enzyme Electrodes	81
6.1	Conclusions from Equations to Predict Tuning Settings	105
6.2	Process Types for Biochemical Loops	106

Foreword

I have watched a company struggle to become a biochemical force of the future through the commercialization of recombinant DNA technology. I have sat through countless seminars designed to teach me how to evolve into a biochemical process control engineer. Unfortunately, nothing was ever said about biochemical measurement and control. Furthermore, I could not find any books on the subject. What was taught in the seminars was what was available in the open literature, which was the principles of biochemistry and fermentor design. Since these facts had little to do with the actual job at hand, they were easily forgotten and relegated to a stack of reference books on my shelf. Everything I learned of importance to me as I functioned in this emerging industry was the result of actual applications and discussions with suppliers and contractors to the biochemical industry. What I learned is summarized in this book, with the hope that it will lighten the load of those instrument and control engineers who wish to make the same transition. There is no information included on cell chemistry. The information on biochemical processes is cursory only. Some of the statistics will be outdated by the time the book is published. Its only intent is to impress the reader with the potential impact of recombinant DNA technology to provide motivation and a framework for learning how the job of the instrument and control engineer will change as the technologies continue to expand and develop.

Gregory K. McMillan
St. Louis, Missouri, 1987

Acknowledgments

The author wishes to express his gratitude to Len Aynardi for information on the state of the art for broth and off-gas composition measurements, Doug Inloes and Gordan Tong for information on the operation requirements of loop fermentors, Stan Weiner for information on the practical considerations in the selection and installation of the field instruments, Paul Sykes for information on the interface requirements for computer systems, and KSHE Sweetmeats for the inspiration.

Biochemical Measurement and Control

CHAPTER 1

An Industrial Perspective

What are biochemical measurement and control? They are the industrial measurement techniques and process control strategies used to get living cells or parts of living cells to produce commercial quantities of useful substances. Each living cell is a sophisticated catalytic reactor that can convert nutrients and sugar into products, as shown in Figure 1.1. The enzymes act as specific catalysts, and the nucleus serves as a control center with the blueprints for chemically complex products.

Why all the fuss about biochemical production? How did it happen that GenenTech® (a genetic engineering firm) set a record

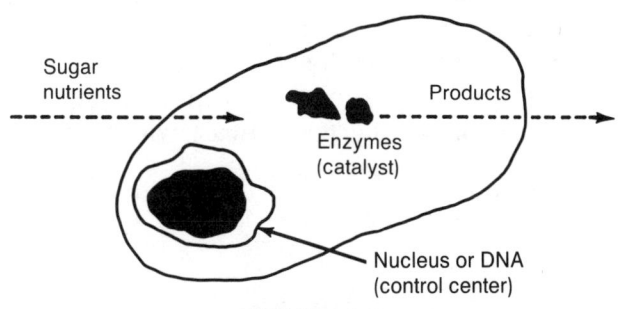

FIGURE 1.1
The Living Cell as a Catalytic Reactor

(1)

for the fastest price rise per share by increasing from $35 to $89 in 20 minutes? We are on the verge of harnessing the ultimate technology, the life process, to provide products to help solve the world's problems. Besides the global benefits of longer life and improved quality of life, there are some local benefits of significant importance to industry in view of its current problems.

Industry has suddenly found itself overwhelmed with competition in the commodity product market. There is a glut of cheap commodity chemicals produced by mature processes. The commodity intermediates (e.g., ethylene) tend to be produced close to the source of cheap feedstocks (e.g., Middle East), while commodity finished goods (e.g., textiles) tend to be produced close to the source of cheap labor (e.g., Far East). It is almost impossible for the West to compete with underdeveloped countries where the wages are a few dollars per hour or with oil-rich countries where natural gas is essentially free (it used to be flared). The United States has moved from a role as supplier of products to the Third World to the role of buyer of products from the Third World. While the economy has been temporarily buoyed by the increase in services (marketing, distributing, advertising, and accounting) for the increase in foreign products, the long-term effect of the continual outflow of money for foreign goods cannot be anything but disastrous. An economy based on local services and foreign industrial production is doomed to failure.

American industry would like to capitalize on its leadership position in basic research and technology development to devise processes for new products. These would generally be high value-added products priced at dollars per gram instead of cents per kilogram like commodity products. The competition would be limited by patent protection and the profit margins would be large. Biochemical specialty products fit this mold. The goal is to learn how to monitor and control the most technically advanced reactor system known — the living cell — to produce incredibly pure quantities of incredibly complex chemical compounds. Some of these compounds as pharmaceuticals will bear the price of several hundred dollars per gram.

1.1
Products

The recent progress in the understanding of the body's defense mechanisms and how cells become vulnerable to attack or malfunction has become a springboard for developing drugs that enhance the body's own ability to prevent or cure diseases. Table 1.1 summarizes the low and high estimates of the annual bulk sales of biotechnology products by the year 2000. While the high estimates listed may be optimistic for chemical, agricultural, and feed products, the high estimate is probably too conservative for medical products in light of the barrage of recent discoveries in the pharmaceutical field (Humphrey, 1984). Within the human body, more and more compounds that control the body's internal response to neurological and physiological disorders are being isolated.

TABLE 1.1
Projected Annual Bulk Sales of Biotechnology Products by the Year 2000

Product	Low Estimate, Billions	High Estimate, Billions
Medical	7	45
Special chemical	5	25
Agricultural	3	9
Nutrition	3	4
Equipment and systems engineering	10	24

Since the greatest potential lies in the medical field, let's take a closer look at some of the reasons for the optimism regarding new drugs. The human body has the best weapons against diseases. When this system malfunctions, as with the disorder known as the Auto Immune Deficiency Syndrome (AIDS), the pharmaceutical substitutes to date are inadequate. Common health problems develop when the body's natural defenses are overloaded due to inadequate production of the necessary compounds or overexposure to the offending foreign substances. What better drug is there than that developed by the human body through millions of years of evolution? Side effects are significantly less because the compounds occur naturally within the body. These "miracle drugs" will have a greater impact on health care than did the development of antibiotics. Such "drugs" can only be produced biochemically because they are too chemically complex to be produced by synthetic methods.

Present biochemical production and purification methods are relatively expensive with low throughput capacity, but the dosage requirements are so small and the health benefits so large that present methods can meet market requirements and still provide large profit margins. The incentive is enough for pharmaceutical firms to presently invest one billion dollars annually in biochemical research. The enthusiasm of the pharmaceutical industry is reflected in the statement by Herbert Weissbach, director of Hoffman-La Roche, Inc.'s Roche Institute of Molecular Biology: "Our base of information is rising exponentially. What we've seen in the past will be a drop in the bucket. Once you know how a normal cell functions and what goes wrong, the possibilities are enormous." (Hall, et al., 1985) Let's look at some of the possibilities.

Of great concern today is a cure for cancer. Biotechnology has discovered several promising methods of cancer therapy. Interferon, a substance secreted by the body's immune system to suppress tumor growth, was the first drug that received great publicity as a cancer cure. By itself it has proven to be somewhat of a disappointment, although it has recently been discovered that it prevents the common cold. Scientists have more recently isolated a tumor necrosis factor (TNF) that causes cancer cells to burst. The search for TNF began after doctors observed that cancer patients would improve remarkably after a bout with a serious infection. Among the body's responses to the infection was the production of the substance TNF that destroys cells. Antibody secreting cells have also been fused to secrete ultrapure or "monoclonal" antibodies that seek out and bind to specific targets. They can be used to directly attack diseased cells or deliver payloads of radioactive or chemical bullets. "Monoclonal" antibodies coupled with interferon and TNF may form a Molotov cocktail powerful and specific enough to deliver the knockout blow to cancer cells (Hall, et al., 1985).

The ultimate weapon for cancer would involve learning what genes and their expressions turn an otherwise healthy cell into a cancer cell. It is suspected that all cells have the potential of becoming cancer cells. The goal is to prevent the formation of the cancer cells in the first place instead of the killing of them afterwards.

The atomic structure of one of the common cold viruses has been mapped. The map shows sites on the virus coat where antibodies can attach to neutralize the virus. Next in line is the AIDS virus. Scientists could develop a generic vaccine that would block sites where the virus passes into the cell.

There are many more disorders whose days are numbered. Table 1.2 lists some of these disorders and the types of biochemical drug being tested for treatment. These drugs are generally more effective and safer than their predecessors.

TABLE 1.2
New Biochemical Drugs Being Tested

Disorder	Drug	Mechanism
Gram negative bacteria	Monoclonal antibodies	Destruction of bacteria by antibodies
Multiple sclerosis	Monoclonal antibodies	Destruction of aberrant white blood cells
Meningitis, herpes, malaria, gonorrhea, croup, strep throat	Vaccine	Vaccine is antigens instead of killed virus or bacteria to trigger antibody production
Hemophilia	Factor VIII	Factor normally present to promote blood clotting
Strokes	Plasminogen-activators	Protein normally present to dissolve blood clots
Arteriosclerosis	Enzyme blocker	Compound to block enzyme that forms cholesterol
High blood pressure	Atrial natriuretic factor	Natural hormone from heart to regulate blood pressure
Osteoporosis	Calcitonin	Natural hormone to regulate calcium levels on blood
Depression, anorexia nervosa, alcoholism	Hormone blocker	Compound blocks hormone in brain associated with disorder
Ulcers	Hormones or hormone-like substances	Natural compound to stop gastric juice secretion and to heal stomach lining

Already a substantial number of pharmaceuticals are produced by biochemical methods. Penicillins and other antibiotics have the largest market value. Over 1000 antibiotics are produced by fungi and over 3000 are produced by bacteria. In 1978 the worldwide sales of antibiotics totaled more than four billion dollars. Vitamins have the next largest market value with worldwide sales of over six hundred million dollars in 1978 (Aharonowitz, 1981).

Some of our most popular foods and drinks are produced biochemically. Contrary to public opinion, the first wine and cheese party was not held in San Francisco. It probably occurred several thousand years earlier on the shores of the Mediterranean when some farmer ate some old moldy chunks of milk and drank some old fermented grape juice. The farmer was joined by friends when the farmer started to dance and talk about stock options. Besides wine and cheese, beer, buttermilk, yogurt, oriental food products such as soy sauce and tofu, pickles, and sauerkraut are produced biochemically by either yeast, bacteria, or mold. More recently, a single-cell protein called Pruteen™ has been manufactured by growing a bacteria that utilizes methanol as a carbon source and energy source. About 75,000 tons per year of Pruteen are produced for animal feed.

In the specialty chemical industry, enzymes, solvents, organic acids, and amino acids are produced biochemically. Enzymes are the compounds within the cell that act as catalysts in the making and breaking of chemical bonds to supply the chemicals needed for cell growth and maintenance. These same enzymes have been commercially used outside of the cell to decompose large molecules such as carbohydrates and proteins. For example, three enzymes are used to convert starch to high fructose corn syrup for soft drinks and three enzymes are used to convert alkenes to alkene oxides for plastics. The enzymatic production of alkene oxides is more flexible and in some ways less expensive than existing synthetic production methods. By changing the feed on which the enzyme acts, different alkene oxides such as propylene oxide for plastic polypropylene and ethylene oxide for plastic polyethylene can be created by the same enzymes. Also, halogen ions are much less expensive to incorporate into the product because a simple salt such as sodium chloride can be used instead of an elemental halogen.

The need for premium unleaded gasoline has increased the demand for ethanol as an octane booster. Ethanol has other uses as a solvent, extractant, antifreeze, and a feed for the synthesis of other organic compounds. Until recently, 70 percent of all the ethanol was made by chemical synthesis. The increased cost of petroleum and the decreased cost of corn has made the biochemical production of ethanol more attractive. The ethanol is secreted by a yeast when fed either a crude sugar or a starch converted to sugar. In some countries such as Brazil and South Africa that have no petroleum resources, ethanol is produced in large quantities as a substitute for gasoline.

Biochemical methods of production of optical isomers, e.g., amino acids, produce only the biologically active isomer, whereas traditional production methods produce both the active and the mirror-image inactive isomers. Thus, half of the synthetic production is biologically inactive. Moreover, the separation of the isomers ranges from being difficult to nearly impossible. Two amino acids in large scale production biochemically are lysine, an essential amino acid for animal feed, and glutamic acid, a flavor enhancer used in the form of a salt called monosodium glutamate (MSG).

Sometime in the next 100 years the oil and gas reserves will vanish. The only renewable source of energy and feedstocks is

AN INDUSTRIAL PERSPECTIVE

biomass. While biomass can be any vegetable mass, the most likely source would be wood chips and wood waste from the wood, paper, and pulp industry. It is technically feasible to produce a synthesis gas for ammonia and methanol from a steam gasification of biomass. Figure 1.2 shows the large number of intermediate chemicals that can be synthetically produced from ammonia and methanol (Baker, 1984).

Biotechnology will affect every aspect of our lives. The biochemical production of pharmaceuticals, food and drink, specialty chemicals, and feedstocks is important, but only part of the story.

Biotechnology promises to have a tremendous impact on agriculture. Bacteria can be developed to fix nitrogen from the air to reduce the amount of fertilizer needed and to secrete a substance that is toxic to insects to reduce the amount of pesticides needed. The crops themselves can be genetically engineered to be more drought- and pest-resistant.

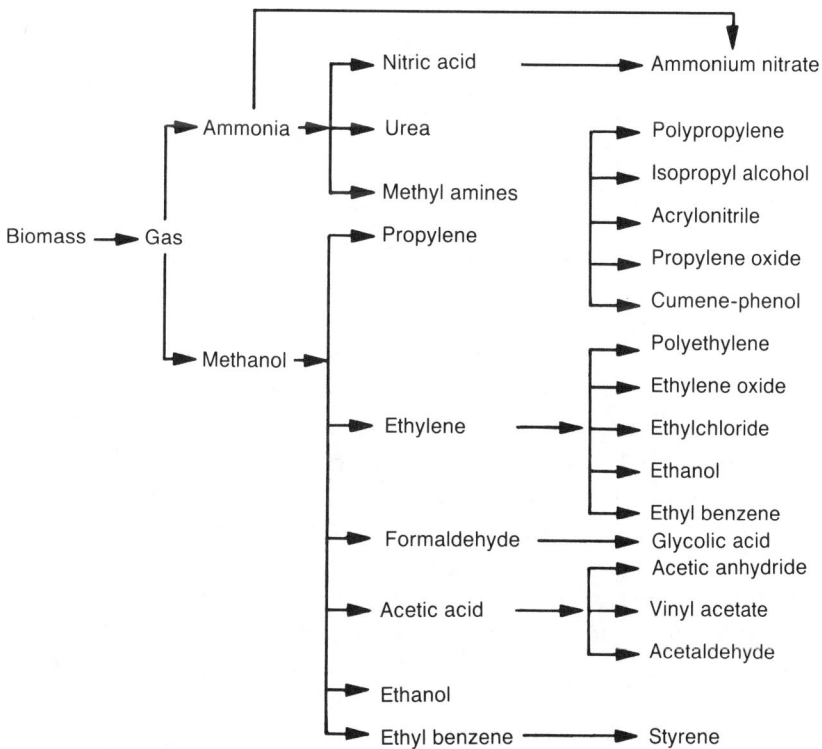

FIGURE 1.2
Chemical Intermediates from Biomass by Gasification

Far into the future, biochips may be replacing microchips to form biological computers that are smaller, faster, and more powerful than anything available today.

How large a role biotechnology has in the future was summed up as follows by Howard A. Schneiderman, a senior vice president of Monsanto®: "Just try to imagine the earth a thousand years from now — in the year 2984. Steam power, electric power generation, nuclear power, transistors, microprocessors, all these will have become historical novelties. Remaining, however, as an indelible part of that future society will be the application and processes of biotechnology. Humanity, using nature's own methods, will have learned to persuade nature to be a full partner in humanity's major enterprise, civilization."

1.2
Process Characteristics

The sudden surge in technological advances in biochemical processes is due to designer genes. I am not talking about the expensive slacks with fancy labels your kids are buying. I am referring to the isolation of genes within a cell that determine the production of a particular chemical, the removal of these genes from one cell, and the relocation of these genes to another cell. The circular DNA double helix that is the genetic material of the recipient cell is broken, the gene from the donor cell is inserted, and the ends of the circular DNA are glued back together (annealed). The new DNA is then placed (transformed) into the recipient cell. The transformed cell then replicates into new cells with the tailored genetic characteristics (Bailey and Ollis, 1977). The whole procedure, as shown in Figure 1.3, takes place in a laboratory. This technology is known as recombinant DNA, or more commonly as genetic engineering. The transformed cell is usually a very simple cell that can be reproduced easily in mass quantities. For example, the genes from a complex mammalian cell that trigger and control the production of a growth hormone can be spliced into the genetic material of a simple bacterial cell. These bacterial cells can survive agitation and can double in concentration every hour, whereas mammalian cells require tepid conditions and several days to double in concentration. The benevolent E coli bacteria found in the human intestine is the workhorse of recombinant DNA because its genetic structure is

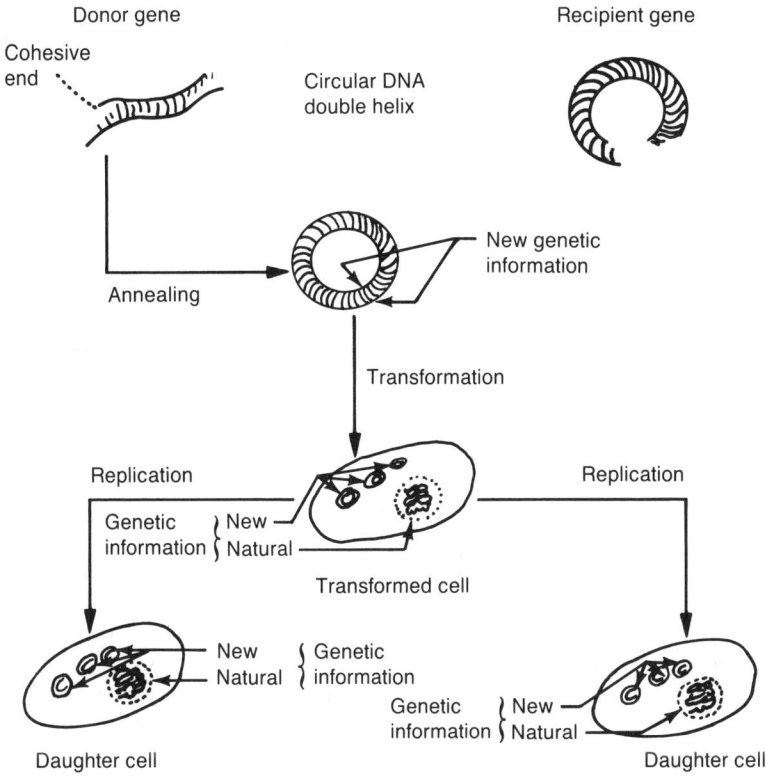

FIGURE 1.3
Recombinant DNA (Genetic Engineering)

simpler and better understood. While simpler in comparison to other cells, it is a reaction system much more sophisticated than anything produced by man. Biochemical engineering seeks to understand and apply the elegant system designs of living cells to the betterment of human civilization. If you had a choice, wouldn't you prefer a reaction system designed by Nature to one designed by your best chemical engineer?

Before we get all starry-eyed with the glorious future of the biochemical industry, we need to recognize some difficult operational and economic problems that must be surmounted.

Living cells require a large variety of nutrients to grow or even just to survive. The normal environment of the cells is complex enough to supply these needs. The movement of these cells to a controlled environment places the burden of responsibility of cell nourishment on the biochemical engineer. A complex mixture of starch or sugar, minerals, and vitamins must be consistently available in just the right concentration. The largest feedstock is starch or sugar. Unfortunately, the price of these materials is subject to large fluctuations, making the economic evaluation of biochemical versus chemical synthetic routes difficult. Just when the cost of sugar compared to a petroleum-based feedstock starts to look attractive, darned if the price of sugar doesn't skyrocket. Alternate sources of energy for cell growth that are cheaper and more abundant, such as wood waste, need to be used. Presently, most cells gag on wood as food, but genetic engineering can increase the cell's appetite for it. Feed cost is a minimum problem for biochemical production methods that must compete with synthetic chemical production methods.

Biochemical products present some challenging separation problems. Cells don't like to live in a too-crowded environment. Just as in a highly populated city, toxic wastes start to accumulate, and the inhabitants have a hard time getting enough fresh air and food. Consequently, the concentration of product from the fermentor is extremely dilute. Until recently, the maximum dry cell weight in the fermentor broth was about one percent or ten grams per liter. By genetically engineering the cells to tolerate higher cell concentrations and biochemically engineering the fermentors to facilitate higher oxygen and heat transfer rates, the dry cell weight can be pushed towards four percent or 40 grams per liter. Since typically less than ten percent of the dry cell weight is product, the product concentration still doesn't exceed four tenths of one percent by weight or four grams per liter in the fermentor broth. Some products are secreted by the cells, but most are trapped inside. Thus, large amounts of water must be removed gently enough not to damage the cells (this generally precludes most distillation methods), and the cells must be carefully broken open and the product removed without threatening the three-dimensional integrity of the complex product molecule.

As if all this weren't challenge enough, add the requirement of sterile operation. The genetically engineered bacteria are generally weaker than normal bacteria and are less resistant to attack from other microorganisms. Bacteria can catch colds and get the flu just like humans, but the results are more dramatic. Viruses known as phages are inert particles until they attach themselves to a bacterium cell as parasites. The phages direct the ribosomes inside the bacteria to destroy the bacteria's DNA and multiply the phage DNA to produce new phage particles within the bacteria. The bacteria cell wall eventually bursts to release new phage particles. In a matter of hours, a fermentor batch can change from a broth rich in bacteria to one rich in phages and broken cells (Bailey and Ollis, 1977).

It is difficult to destroy these extremely small inert particles. After a phage contamination in a production system, it may take several weeks of formaldehyde washing of all equipment, piping, connections, and vents before production can be resumed. The probability of subsequent contaminations is great after the first contamination. Nothing can ruin a plant manager's career faster than a phage contamination.

Even if another microorganism doesn't attack the bacteria, it can compete for the food and air supply and become a contaminant that is difficult to separate out downstream. Some bacteria can form endospores when environmental conditions are not to their liking. In this dormant state they can resist heat, radiation, and poisonous chemicals. When the surroundings improve, they become active and germinate to create normal functioning cells. Thus, sterilization of the feeds doesn't guarantee that a fermentor will not contain other bacteria. Bacteria exist everywhere. Billions thrive on the human skin. The average hamburger has 5 to 10 million per gram.

Ultrafiltration and clean rooms are used in conjunction with sterilization to prevent the introduction of bacteria into the biochemical process or product.

For products that are going to be injected into humans or animals, exceptional care must be taken that no pyrogens have been formed. Pyrogens are toxic byproducts of bacteria that can cause an immediate fever reaction upon injection in quantities as low as nanograms.

The U.S. Department of Health, Education and Welfare published a set of Current Good Manufacturing Practices (CGMP) for the manufacture of human and veterinary drugs in 1977. Failure to comply results in classification of the drug as adulterated and makes the responsible person subject to regulatory action. The CGMP is quite comprehensive in that it details the requirements for organization and personnel, buildings and facilities, equipment, production controls, packaging controls, holding and distribution, laboratory controls, records and reports, and returned or salvaged drugs.

While the bacteria, fungi, and yeasts used for biochemical production are harmless, once they have been altered genetically, there are Federal guidelines on their confinement. The National

Health Institute (NHI) has detailed the requirements for preventing the release of recombinant DNA microorganisms. At the end of most fermentations, the microorganisms are killed by either high temperature or pH so that no living cells carry over into the downstream process.

Figure 1.4 a functional block diagram of feed preparation and reaction for biochemical processes. Cells need carbon, nitrogen, oxygen, iron, phosphorus, magnesium, and trace elements to grow. Carbon is needed in the largest amount as a source of energy and building blocks for a cell's structural components. It is usually supplied as a sugar such as glucose or as a starch that can be converted to a sugar by saccharifying agents. This raw material is known as the substrate. As with all feeds, it is filtered and sterilized. The cost of this substrate is typically more than 50 percent of the operating cost. Even the manufacture of the single-cell protein Pruteen, which uses a relatively cheap methanol instead of sugar as its carbon source, has a substrate cost that is about half of its operating cost. It is important to try to recover as much of the leftover substrate in the fermentor broth as possible. In practice, this is difficult for batch operation because of the potential cross-contamination of batches from residual microorganisms or their byproducts.

The preparation of the nutrient mix is more an art than a science. It is impossible to predict the exact quantities of the trace nutrients that will optimize the fermentor yield. The amounts depend upon the microorganism, substrate, nutrients, product, fermentor design, and operating conditions. The recipe is usually the result of extensive experience gained from test batches in a research lab or pilot plant. Process computers play an important role in providing repeatable conditions and recording correlated results. Technicians with manual valves and clipboards add too much uncertainty to home in on the optimum mixture.

The oxygen is supplied by sparging either air or oxygen-enriched air into the seed culture vessels and the fermentor. The amount of dissolved oxygen available for transfer to the cells increases with broth temperature and backpressure.

The seed culture vessels are actually fermentors on a much smaller scale. The genetically engineered cells start out as a relatively small number in a laboratory flask. The total number is increased by providing the right conditions for them to multiply and moving them to successively larger vessels.

The biological reaction or conversion occurs in a vessel known as a fermentor. Substrate, nutrients, and oxygen must be supplied

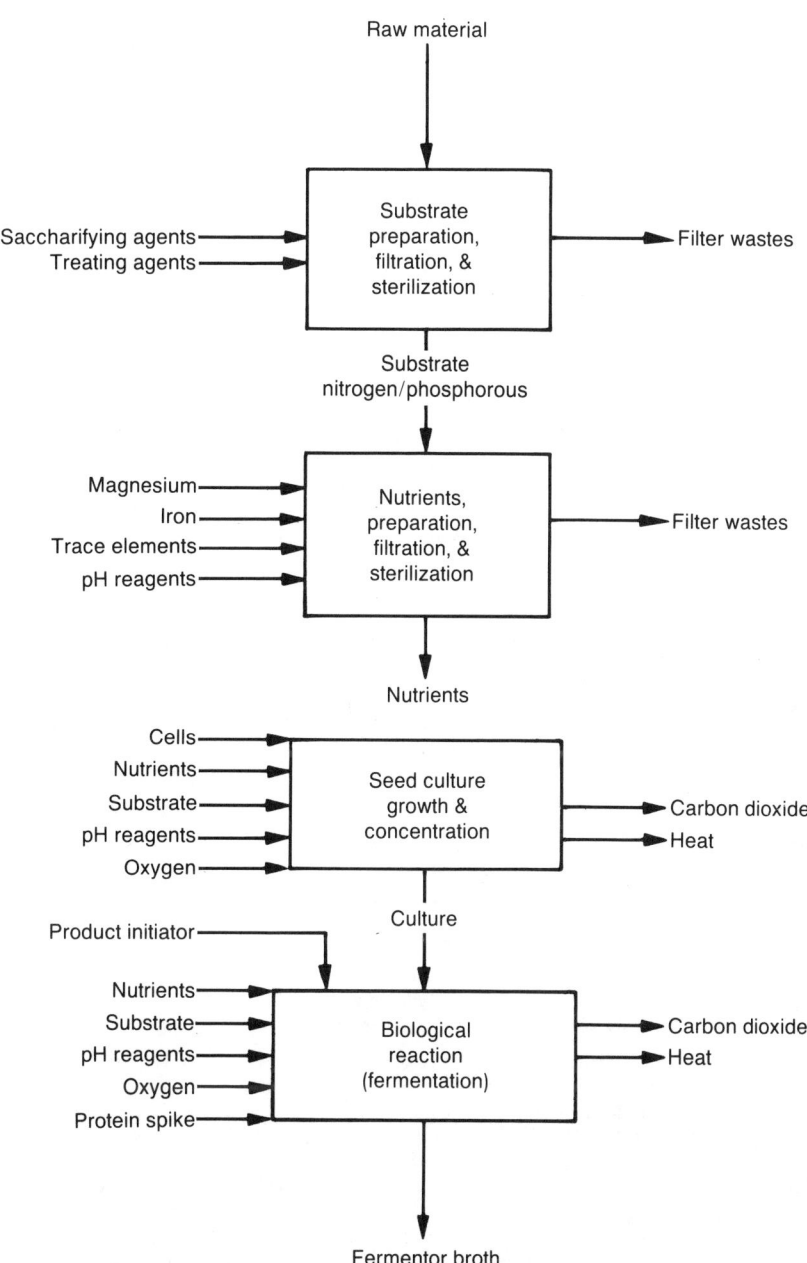

FIGURE 1.4
Feed Preparation and Reaction

in an amount proportional to the number of cells and their growth rate. Too little or too much of any of these feeds can cause the rate of cell growth or product formation to decrease and undesirable side reactions to increase. For example, too high or too low a dissolved oxygen level not only suppresses cell growth but also causes the production of acetate, which must be neutralized by the addition of a reagent such as ammonia to keep the pH constant.

The cell concentration and growth rate vary tremendously during a batch as shown in Figure 1.5. When the seed culture is first placed in the fermentor, the cells require time to become acclimated to their new environment, establish a balance between their chemical components and that already in solution, and produce intermediates required for growth. The result is a lag where the cell growth rate is negligible. The length of the lag phase can be reduced by initially providing some essential amino acids required for growth instead of waiting for them to be produced by the cells. This protein spike is particularly effective for a young culture because inhibiting toxins have not had time to accumulate. The lag phase is followed by an exponential growth phase where the growth rate increases with time. The cell concentration eventually becomes large enough to become an inhibiting factor. The growth rate decreases and the cell concentration reaches a maximum and stays nearly constant in the stationary phase. All living things die. Eventually, the death rate exceeds the birth rate

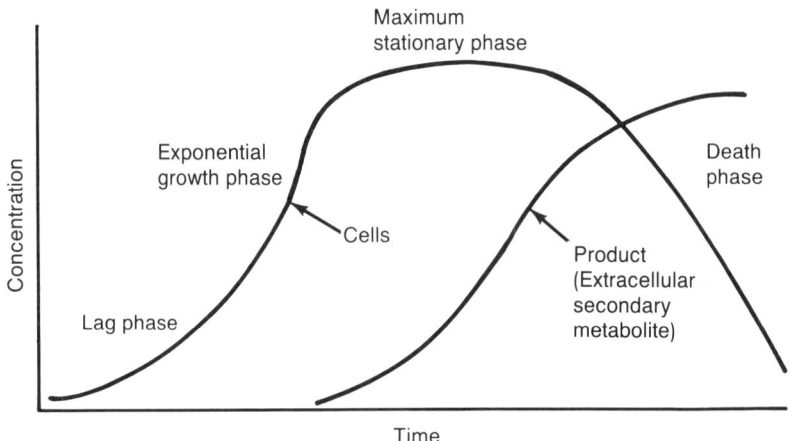

FIGURE 1.5
Batch Cell Growth and Product Formation Curves

due to the depletion of nutrients and the accumulation of toxins. However, the batch is usually killed and the broth transferred to the separation and purification area before the death phase is reached.

The rangeability required of the oxygen flow controller is greater than can be measured or manipulated. Table 1.3 shows the oxygen transfer rates required for different times and cell concentrations for E coli bacteria. At the beginning of the batch, the air flow rate cannot be turned down low enough and the dissolved oxygen level floats way above set point, close to 100 percent of saturation. At the end of the batch, even if the air flow was large enough, the oxygen transfer rate of conventional agitated fermentors is inadequate, the dissolved oxygen level drops toward zero, and oxygen starvation occurs. Consequently, jet fermentors have been recently designed to increase the oxygen transfer rate. There is a similar but somewhat lesser rangeability problem with the substrate, nutrient, and reagent addition if these are to be kept at a constant concentration in the broth throughout the batch.

TABLE 1.3
Glucose and Oxygen Rangeability Requirements for E coli

Batch time, hours	Cells concentration, gm/liter	Glucose consumption, gm/(hr·liter)	Oxygen consumption, mm/(hr·liter)
0	0.1	0	0
2	0.5	1.0	12.5
4	2.6	5.1	65.0
6	38.7	75.5	967.0

If the product is the cells themselves (cellular product), as is the case with the single-cell protein Pruteen, the cell growth curve is the product formation curve. More often the product is secreted by the cells (extracellular product) or trapped inside the cells (intracellular product). A special nutrient called a product initiator is used near the end of the exponential growth phase to start the formation of extracellular and intracellular products. The product formation curve lags the cell growth curve and peaks near the end of the stationary phase.

Once the product has been formed in the broth, the arduous task of separation and purification begins. Figure 1.6 shows the functional blocks for the separation and purification of cellular, extracellular, and intracellular products. Most of the equipment

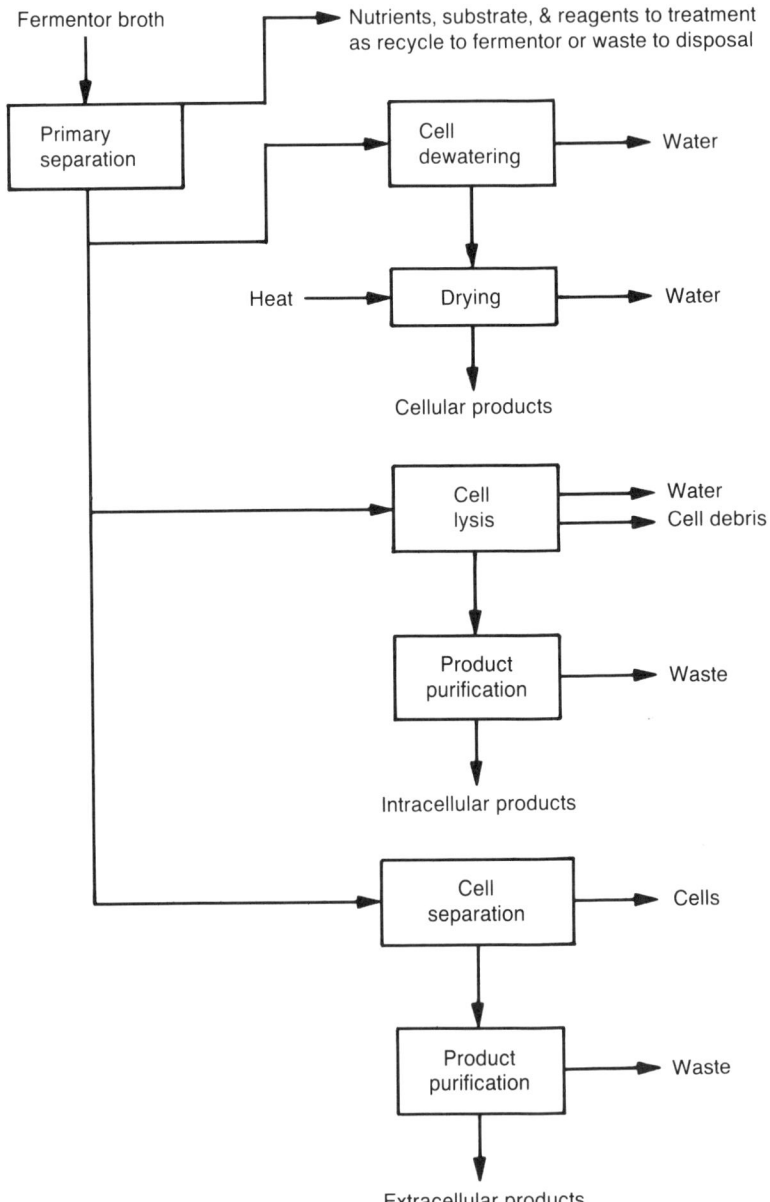

FIGURE 1.6
Product Separation and Purification
(Cellular, Intracellular, and Extracellular Products)

cost and operating cost outside the substrate cost are in this downstream processing.

The primary separation removes the leftover nutrients, substrate, and reagents. These concentrations are never zero in the fermentor broth because they are required for normal maintenance activities independent of whether the cells are still growing. Since the product concentration is less than one percent, large amounts of liquids must be removed. Techniques such as flocculation, filtration, and centrifugation are used. For cellular products, only further centrifugation for cell dewatering and rotary drying for granule formation are required. The extracellular and intracellular products require methods with greater selectivity such as chromatographic, membrane, and electrokinetic techniques. Many of these techniques have been demonstrated in the laboratory, but the scale-up problems and operational problems (e.g., fouling) to facilitate their use on an industrial scale have largely not been solved. The company with an edge in biochemical processing will be the one to first develop reliable and economical purification techniques for its intracellular and extracellular products. Conventional chemical methods such as distillation are not applicable except for the purification of organic solvents such as ethanol because the higher temperatures cause product degradation.

The waste disposal problems posed are significant due to the large volume of liquids handled and confinement requirements for recombinant DNA organisms. Any attempt made to increase the concentration of the product in the fermentor broth is well worth the effort because of the reduction in waste treatment costs.

Continuous operation is more efficient for autocatalytic reactions where the conversion rate increases with concentration. A biochemical reaction is autocatalytic during the exponential growth phase. Thus, a continuous stirred reactor (CSTR) for this phase followed by a plug flow reactor (PFTR) for the stationary phase would be the most efficient mode of operation for most biochemical reactions. However, in practice nearly all biochemical reactions are fed batch, where some of the advantages of continuous operation are achieved in a batch mode. In fed batch operation, the substrate and nutrients are not all charged at the start of the batch but are fed at varying rates throughout most of the batch in accordance with their depletion rates. There are many practical reasons why continuous biochemical reactors have not been used. The most notable exception is the continuous fermentor for the production of the single-cell protein Pruteen. A gram negative bacteria is used that is hardier and able to withstand antibiotics used to suppress the growth of other microorganisms. Some of the problems associated with continuous biochemical operation are listed as follows:

(1) Contamination from the accumulation of toxins and microorganisms
(2) Loss of batch identity
(3) Excessive nutrient and substrate cost (larger amounts in effluent)
(4) Inadequate long-term stability of microorganisms
(5) Inadequate knowledge of reaction kinetics
(6) Lack of on-line composition measurements

1.3
Measurement and Control Characteristics

Biochemical processes require sensors that are more accurate and more reliable while being capable of withstanding steam sterilization. The microorganisms used in biochemical reactions are fussy about their environmental conditions. In order for these cells to do their thing (convert substrate into product), the temperature, pH, dissolved oxygen, and dissolved carbon dioxide levels must be held within a few percent of an optimum value. The normal environmental conditions for E coli bacteria is that of the human intestine where the temperature is always a cozy 98.6 degrees and the pH just slightly less than neutral. If one of these sensors fails and causes erratic control, a batch representing six to 96 hours of processing can be ruined. For high value-added products, the cost of production downtime can be enormous. For all products, the cost of the substrate wasted is significant. On top of all this, the sensors must be accurate and reliable after exposure to the high temperatures of repeated sterilizations.

Measurements provide the only window into the biochemical process. If the view is distorted, not only is the data wrong, but the user might not even know it is wrong. This insidious nature is best illustrated by a pH electrode that has become fouled for a pipeline (low residence time) pH control system. The electrode response time increases drastically, but the net effect seen is attenuated but with longer oscillations. The process pH oscillations are larger, but the filtering effect of the slower electrode causes the measured oscillations to be smaller. The pH controller can even use a higher gain. So far as the user is concerned, the pH control looks better when, in fact, it has gotten worse.

The glossy catalog brochures with their impressive specifications give the user a false sense of security as to the performance of the measurements. Experience from industrial applications shows that the sensor is the weakest link in a control system: any sensor can fail to any signal level, every sensor will eventually fail, and the installed accuracy is worse than the specified accuracy. If a measurement becomes slow or inaccurate, it is difficult to detect it on-line before the damage is done to the process unless multiple sensors are installed. Most fermentors now use two pH and two dissolved oxygen (DO) electrodes. The only problem is that when they disagree (which they always do) the user doesn't know which one to believe. Concentration gradients and differences in age, coatings, sterilizations, and velocities guarantee the measurements will disagree. Three pH and three DO electrodes are needed so that the electrode measurement furthest from the other two can be discarded. This is accomplished automatically by the median selector functional block in distributed control systems. The median signal is used as the controlled variable. Upscale or downscale failure of one electrode does not cause a failure of the control system. A deviation alarm is used to warn the user when one of the electrodes deviates from the median by an unacceptably large margin. This forewarning allows diagnostics and troubleshooting to be done in a logical, organized fashion instead of in a panic mode.

The whole purpose of a control system is to minimize the difference between the actual process variable represented by the measurement and the desired value represented by the set point. Measurement noise, measurement error, control error, and final element error all contribute to the total error in the process variable. The secret to better biochemical process control loop performance requires that the user first recognize the relative magnitudes and causes of these errors.

Before you can appreciate the reasons why the control errors for load disturbances are large or small, you need to understand some fundamental concepts about control loop dynamics. The peak control error for any given load disturbance corresponds to the maximum excursion of the process variable during the loop dead time. The loop dead time is that period of time that starts when a disturbance enters the process and continues until the controller has manipulated something that starts to cancel out the effect of the disturbance. The loop dead time is the sum of all transportation delays, mixing delays, heat transfer lags, sample delays, sensor response lags, analyzer or controller computational delays, and final element response lags in the control loop. Before a control loop can stop an excursion, the controller needs to see the effect of the disturbance and manipulate something that cancels out the effect of the disturbance. A useful analogy is the "purple passion" party. The novice party-goer drinks one glass after another of grain alcohol and grape juice during the dead time when he or she doesn't feel the effect of the alcohol and has no inclination to stop since it tastes harmless. The degree of intoxication reached depends upon the number of drinks or the maximum excursion of the alcohol concentration in the blood during this dead time. The dead time depends not only upon physiological parameters, but also upon the control center or brain and its ability to recognize what is happening and to reduce the drinking rate. A runaway or positive feedback condition can develop where the intoxication level becomes great enough for the brain to lose inhibition and increase the drinking rate until an equipment failure occurs (the person passes out). The rate of rise of the process variable (intoxication) depends upon the size and speed of the disturbance (size and frequency of drinks) and the process volume (body size). For a proportional-plus-integral-plus-derivative (PID) controller and a proportional-plus-integral (PI) controller that are properly tuned, the peak control error is the maximum excursion reached after 110 and 150 percent of the loop dead time, respectively, as shown in Figure 1.7 (McMillan, 1983).

Table 1.4 summarizes the sizes and sources of errors for some of the more important types of control loops in the reaction-conversion area and the separation-purification area of biochemical processes. The reaction-conversion area was considered to consist of well-agitated vessels with relatively large volumes, and the separation-purification area was considered to consist of in-line filters, membrane units, and chromatographic columns with relatively small volumes.

For pH and dissolved oxygen (DO) control loops, the process equipment dead time due to the turnover time for pH or the bubble rise time for DO is small in comparison to the residence time for continuous operation and the ramp rate for batch operation in the reaction-conversion area. The dead time from the electrode response lag is larger but the same order of magnitude as the dead time from the process equipment. However, due to the large volumes involved, the process variable excursion rate is slow

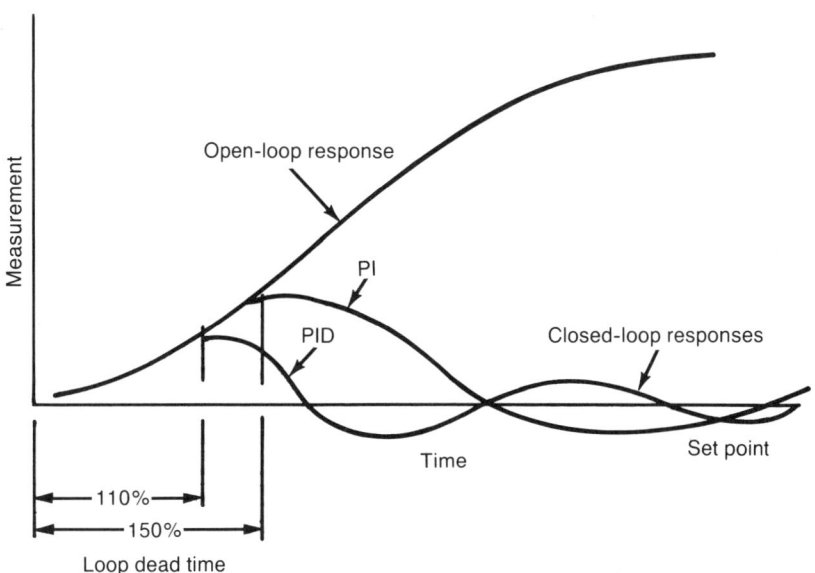

FIGURE 1.7
Peak Control Error for a Load Disturbance

so that the magnitude of the excursion during the total loop dead time is relatively small.

The measurement error for pH and DO is larger than the control error in the reaction-conversion area. Many things can cause electrode errors, as described in Chapters 3 and 4, but electrode coatings or repeated sterilizations are the culprit more often than not. When it comes to measurement noise, pH and DO error sizes differ significantly. Unless the pH measurement circuit is subjected to electromagnetic interference (emi) or radio frequency interference (rfi), pH measurement noise is much smaller than DO measurement noise. Gas bubbles like to alternately attach and release themselves from DO electrodes and cause dramatically different readings due to the difference in oxygen transfer rates between the gas and liquid phases.

For temperature control loops, the control error is also relatively small in the reaction-conversion area. While the absolute magnitude of the process equipment dead time due to heat transfer lag is larger than that for the pH loop, the size of the excursion during this time is still small due to a large vessel volume and small disturbance size. The disturbance size is small because a cascade control loop is used, where the inner or

TABLE 1.4
Relative Magnitudes of Errors in Biochemical Processes

Process variable	Error type	Reaction-conversion		Separation-purification	
		Size	Major sources	Size	Major sources
pH	Control	Small	Turnover time & electrode lag	Large	Transportation & electrode lag
pH	Measurement	Large	Electrode coating & sterilizations	Large	Electrode coating & sterilizations
pH	Noise	Small	Ion fluctuations & gas bubbles	Large	Ion fluctuations & gas bubbles
DO	Control	Small	Bubble rise time & electrode lag	—	—
DO	Measurement	Large	Electrode coating & sterilizations	—	—
DO	Noise	Large	Air fluctuations & gas bubbles	—	—
Temperature	Control	Small	Heat transfer lag & thermowell lag	Medium	Heat transfer lag & thermowell lag
Temperature	Measurement	Small	Nonlinearity & electronic	Small	Nonlinearity & electronic
Temperature	Noise	Small	Heat fluctuations	Small	Heat fluctuations
Composition (liquid)	Control	Large	Growth lag phase & sample delay	Large	Transportation & sample delay
Composition (liquid)	Measurement	Large	Electrode coating, wear, & aging	Large	Electrode coating, wear, & aging
Composition (liquid)	Noise	Medium	Conc. fluctuations & gas bubbles	Medium	Conc. fluctuations & gas bubbles

secondary loop compensates for tempered water disturbances before they can affect the outer or primary loop, which is at vessel temperature.

The measurement error of temperature loops is rather small because resistance temperature detectors (RTD) and narrow calibration spans are used. Measurement noise is virtually nonexistent if there are no emi or rfi problems. The overall accuracy, control, and reliability of temperature loops, if properly designed, installed, and tuned, are greater than those of any other loop. Some common pitfalls cause the actual performance to fall far short of the expected. Too large an air gap between the RTD sheath and the thermowell inner wall causes too large a measurement lag or too large a signal gap between the split ranging of the final elements for steam, and chilled water flow causes too large a final element dead time.

In liquid composition control loops in the reaction-conversion area, even though the process equipment dead time due to mixing is small, the process dead time due to the growth lag phase is very large. When the substrate concentration is increased, the cells do not immediately increase in concentration. It takes time for them to process the food and manufacture the components necessary for replication. In fact, an inverse response exists: after an increase in substrate feed, the initial response is the opposite of the final response. The substrate concentration first increases and the cell concentration decreases due to dilution, and then the substrate concentration decreases and the cell concentration increases due to cell growth. The controller must be detuned and the use of rate or derivative action precluded to avoid overreaction to the initial process response.

The measurement error broth composition loop depends upon the method used, but it is nearly always large. Until recently, cell concentration was inferred from an integration of either the oxygen uptake rate (OUR) or carbon dioxide production rate (CPR). The OUR and CPR were calculated by subtracting the measured oxygen and carbon dioxide concentrations, respectively, in the vessel off-gas from a reference value in the gas feed (usually air). While the use of mass spectrometers greatly improved the accuracy of the off-gas concentration measurements, variations in the oxygen and carbon dioxide transfer rates, uncertainties as to how much respiration is used for growth versus maintenance, and variations in the substrate concentration caused the inferred cell concentration measurement to be inaccurate and noisy. The development of the biosensor to measure glucose, the most common substrate, provides a method of direct substrate concentration control or inferred cell concentration control that is less prone to measurement error or noise. The biosensor is not sterilizable at present. Consequently, it cannot be inserted directly into the vessel like the pH and DO electrodes but requires a sample be withdrawn continuously for on-line control. The sample transportation delay is minimized by reducing the length of sample line between the vessel and the electrode. The long-term accuracy and reliability of the biosensor in industrial applications needs to be established. There are many chromatographic techniques for liquid composition measurements, but these are relegated to an off-line check rather than an on-line control mode. One interesting prospect of cell growth measurement involves pumping ultrasound into a vessel and measuring the amount of signal beamed back. The reflected signal is expected to be an accurate measure of cell growth. The ultrasound emitter and detector can be mounted on the exterior vessel walls so that sterilization is not required. Chemically sensitive field transistors (CHEMFETs), microsensors that consist of semiconductors bonded to a compound whose chemical or physical form changes in the presence of a particular species, promise to revolutionize composition measurements. Their size (about that of a dime) and their cost (probably less than $100) will enable *in situ* redundant measurements that are thrown away once they exceed their useful life (Weiss, 1985).

In the separation-purification area, the control errors are much larger. The smaller volumes mean the rate of change of the process variable for a given disturbance is much larger. In many cases, the process equipment dead time due to transportation time delays is also larger. This is especially true for control loops around chromatograph columns and multiple-pass filter or membrane units. Measurement errors are about the same as in the reaction-conversion area, but pH and composition measurement noise is larger due to composition fluctuations from insufficient backmixing. In general, the performance of control loops in the separation-purification area is less, but the requirements may also be less, because at this point the cells have been killed and the precision of environmental conditions can be relaxed.

The fermentors are the most highly instrumented pieces of equipment in biochemical processes. Figures 1.8 and 1.9 show the instruments typically found on stirred and loop fermentors. The loop fermentor has been developed to meet the greater oxygen transfer and heat transfer requirements of the higher cell concen-

AN INDUSTRIAL PERSPECTIVE (25)

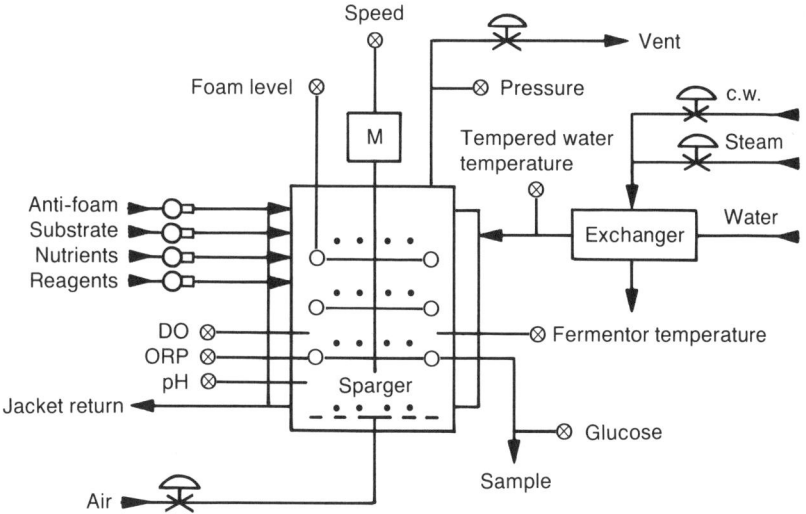

FIGURE 1.8
Stirred Fermentor Measurement and Control

trations approached in new processes via genetic engineering of the cells. The jet action provides more gas bubbles and turbulence, and the external heat exchanger provides larger heat transfer areas and heat transfer coefficients.

For the same control loop configuration, the pH control error is larger for the loop fermentor because the turnover time is three to four times larger. If the pH electrodes are located in a recirculation line, the high fluid velocity (15 fps) reduces the electrode response lag enough to offset the increase in process equipment dead time — enough to keep the control error about the same as for the stirred fermentor. The high velocity also dramatically reduces the incidence of coating problems. The fluid flow streams must not directly impinge on the electrode tip; otherwise, the electrode life will be shortened. The amount of measurement noise for the pH and DO electrodes is much less in the recirculation line than in the fermentor because of fewer gas bubbles.

For the same control loop configuration, the temperature control error will also be slightly larger for the loop fermentor due to the poorer mixing, the additional transportation delays from the recirculation loop, and fouling of the additional exchanger in the recirculation loop. The last two problems can be eliminated

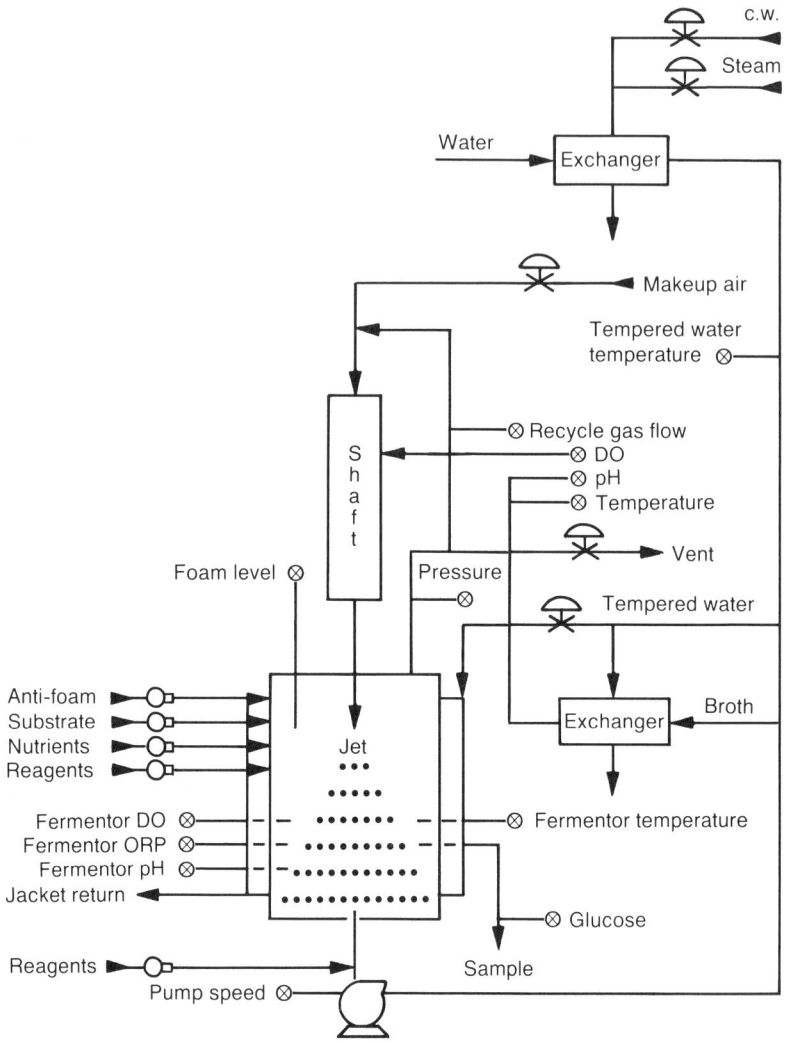

FIGURE 1.9
Loop Fermentor Measurement and Control

by the use of a recirculation temperature control loop to manipulate the set point of the tempered water loop. The tempered water exchanger and control system must be retained because the high or low wall temperatures in the recirculation exchanger from steam or chilled water would tend to locally shock the cells in the recirculation loop.

Local oxygen depletion and low pH excursions in the recirculation line will occur. The DO and pH reaction times are in the order of seconds, whereas the residence time in the recirculation line approaches one minute. Also, over one third of the working volume is in this line. Consequently, DO and pH electrodes should be installed in the recirculation line downstream of the exchanger. The addition of a cascade control loop, where fermentor pH is the outer (master) loop and recirculation line pH is the inner (secondary) loop, sounds good in principle to improve fermentor control, until you realize that the same strict limits on environmental conditions for the cells in the fermentor also apply to the cells in the recirculation line. Subsequently, there is no freedom to manipulate an inner loop set point. This same problem occurs for temperature and DO control. DO cascade control is not feasible for another reason. The use of air flow to the shaft as the manipulated variable for DO control is downstream of the recirculation line measurement so that it affects fermentor DO measurement before it affects the recirculation line DO measurement.

Given that you must control the recirculation line process conditions, the question is then "What do you do with the fermentor measurements?" One alternative is to use the measurements of fermentor temperature, pH, and DO as indicators of fermentor conditions and to use the measurements of temperature, pH, and DO installed downstream of the recirculation line exchanger as the controlled variables for control of recirculation line conditions. The temperature controller manipulates the tempered water temperature, the pH controller manipulates reagent injection into the suction of the recirculation line pump, and the DO controller manipulates recirculation pump speed. The air flow to the shaft or the fermentor backpressure is manipulated very slowly by an optimizing controller to keep the pump speed in the usable range. If the pump speed is too low, the aeration in the fermentor from the jet is insufficient. If the pump speed is too high, energy is wasted. The response of the DO loop to speed changes is very fast and deals effectively with short-term transients. The air flow or backpressure is manipulated to handle slower transients from cell growth. Any change in recirculation line pump speed is used as a feedforward signal to change the output of the temperature and pH controllers to decouple these loops from the DO loop. Another alternative is to use the fermentor measurements to form a second set of control loops. A fermentor temperature controller would manipulate tempered water flow to the jacket; a fermentor pH controller would manipulate reagents added directly to the fermentor; and a fermentor DO controller would manipulate the air flow or backpressure. The fermentor control loops would interact with the recirculation line loops. Hopefully, the faster response of the recirculation line loops would minimize the detrimental effects of interaction. For the high recirculation rates expected for loop fermentors (two to three turnovers per minute), it appears that the first alternative is adequate even for the startup mode when the greatest difference in operating conditions between the fermentor and the recirculation line occur. Table 1.5 summarizes the control loops for stirred and loop fermentors.

TABLE 1.5
Typical Stirred and Loop Fermentor Control Loops

Measured variable	Location	Manipulated variable(s)
Stirred fermentor		
Fermentor foam level	Fermentor top	Anti-foam agent pump
Fermentor backpressure	Fermentor top	Vent valve
Fermentor temperature	Fermentor side	Tempered water temperature (to fermentor jacket or coil)
Tempered water temperature	Exchanger outlet	Steam & chilled water valves
Fermentor pH	Fermentor side	Acid & base reagent pumps
Fermentor DO	Fermentor side	Air valve, agitator speed, & fermentor backpressure
Fermentor glucose conc.	Sample line	Substrate pump
Loop fermentor (besides above loops)		
Recirc. line temperature	Exchanger outlet	Tempered water temperature (to recirc. line exchanger)
Recirc. line pH	Exchanger outlet	Acid & base reagent pumps
Recirc. line DO	Exchanger outlet	Recirc. line pump speed
Recycle line gas flow	Valve inlet	Recycle gas valve

Biochemical pH control has some attributes that, while subtle, are important in getting a proper perspective of the control problem. Cells have their own pH control systems. You cannot force the internal pH of the cell to be a certain value by regulating the external pH. Thus, at first glance pH control may seem superfluous. However, there is documented proof of how the pH of the broth drastically affects the cell growth rate. The apparent contradiction can be resolved when it is recognized that adverse external conditions can cause the internal mechanisms of the cell to falter. What causes the pH of the broth to vary and warrant a control system? Most of the pH changes are negative and are due to the production of acetate by the cells whenever the dissolved oxygen level is either too high or too low (this assumes that too much ammonia was not added as a nutrient). Consequently, tighter DO control could reduce the need for pH control in the fermentor. Even if this is achieved, there will still be lots of pH problems in biochemical plants due to needs in the feed preparation, separation/purification, and waste treatment areas.

The greatest opportunity for the application of advanced control algorithms such as the model predictive controllers (e.g., dynamic matrix controller and forward modeling controller) is in the area of broth composition control. The ultimate goal is to control and optimize the product concentration. For continuous fermentations, this corresponds to a single-point optimization problem, but for fed batch fermentations, this corresponds to a time optimal or path profile optimization problem. In simpler words, a single optimum product concentration set point is needed for continuous operation, whereas a product concentration set point as a function of batch time is needed for batch operation. The model predictive controllers perform particularly well for the concentration set point changes of fed batch operation because the large dead time associated with cell growth can be compensated for in the set point response of the model. Even more importantly, the tuning settings of these algorithms can be calculated more accurately so that they can be altered as the batch progresses and the process dynamics change. There is one hitch in this whole idea. Most growth processes exhibit a nonselfregulating response during the exponential growth stage. The rate of increase in cell concentration increases as the cell concentration increases. If the response looks like an integrator, the use of a very large time constant in the process model for the advanced algorithm may be sufficient to insure stability. If the response looks like a runaway, there is no method to adequately include a positive feedback time constant in the process model. Thus, for biochemical processes that show strong autocatalytic characteristics, it is best to revert back to conventional PID control of product concentration (McMillan, 1983).

Some studies get carried away and put advanced control algorithms on all the fermentor control loops. These algorithms perform no better than well-tuned conventional PID controllers for load disturbances to the temperature, pressure, pH, and DO loops (McMillan, 1986). While these algorithms may perform slightly better for set point changes, the control errors are already smaller than the measurement errors. One's time and money are better spent on improving the performance of the measurements than on implementing advanced control algorithms.

Finally, there is the question of what the final element is for biochemical control. Any final element that directly controls a fluid in contact with the process stream must be both sanitary and steam-sterilizable. There must be no crevices, pockets, or porous surfaces to trap bacteria; no packings, glands, or bearings to contaminate the fluid; and no deterioration from the higher temperatures of steam sterilization. Diaphragm valves and diaphragm pumps have been used to date, but these typically do not have the rangeability required for pH, DO, and broth composition control. Also, the capacity of diaphragm valves is too large for the manipulation of reagents, nutrients, or substrates for all pilot plants and most industrial plants, and the pulsation from diaphragm pumps creates excessive measurement noise for in-line composition control in the separation/purification area. Recently, H. D. Baumann Assoc., Ltd., has developed a packless control valve that can manipulate extremely small flows with a precision and rangeability that is an order of magnitude better than that for diaphragm valves or diaphragm pumps.

While most of the streams to be manipulated are small additive flows, sometimes it is desirable to manipulate recirculation flows. For a loop fermentor, the recirculation flow is in the thousands of gpm. If a sanitary and sterilizable control valve were available in large sizes, bypass control of the heat exchanger in the recirculation could be used to provide faster temperature control. The blending of the streams that pass through and around the exchanger gives a faster temperature response than the manipulation of the temperature of the tempered water. As long as the temperature of the stream that passes through the exchanger is kept just a few degrees below the set point, the decrease in cell growth rate and the increase in energy cost are small enough to warrant the better temperature control. In the food and drink industry, Teflon®-lined butterfly valves have been used for such high flow rate applications. However, these valves do not strictly meet the sanitary requirements because they have flanged connections and stem packing. There is no way such valves could be used in the manufacture of products by recombinant DNA microorganisms.

CHAPTER 2

Flow, Level, Pressure, and Temperature Measurements

Most of the biochemical process control loops involve flow, level, pressure, and temperature measurements. Flow measurements are used on all nutrient, reagent, substrate, and gas feeds in the reaction/conversion area and on many throughput, recycle, and waste streams in the separation/purification area. Level or inventory measurements are used for level control loops in continuous vessels and for charge totalization in batch vessels. Pressure measurements are used for the vapor space pressure control of equipment so that a positive pressure is always maintained to prevent infiltration of microorganisms from the environment. Temperature measurements are used throughout biochemical processes to maintain the precise temperature conditions required by cells and their products.

In general, the performance of these instruments profoundly affects the total error of their respective control loops. The measurement error is frequently larger than the average control error. The response time constant of each of these measurements

(31)

is usually longer than the process dead time so that the measurement significantly increases the control error as well. For flow, level, and pressure measurements, the response time constant is also larger than the process time constant. The result is a particularly devious effect on the apparent control error. As the response time constant is increased, the measured control errors decrease, but the actual control errors increase. The measurement time constant slows down the ability of the controller to react to disturbances (hence, the actual control errors increase) but filters the process oscillations (hence, the measured control errors decrease) (McMillan, 1983).

Herein lies a notable difference between biochemical and chemical control loops. Most chemical control loops use relatively large control valves. Consequently, the size of the pneumatic actuators makes the control valve the slowest element in a flow, level, or pressure control loop. The control error of these loops is affected much more by the speed of the valve than by the speed of the measurement (McMillan, 1985). Most biochemical control loops use small control valves or metering pumps. Now, the importance of the measurement response time approaches and sometimes exceeds the importance of the final element response time.

Any sensor that is in contact with a process fluid must meet both sanitation and sterilization requirements. There must be no crevices, pockets, or porous surfaces to trap bacteria; no packing, glands, or bearings to contaminate the process; and no deterioration from the higher temperatures of steam sterilization. Smooth and polished stainless steel surfaces and Tri-Clover™ end connections are generally used in short batch operations (Tri-Clover is a trademark of the Ladish Company).

2.1
Flow Measurements

Flow measurements must be of the in-line type to be sanitary and sterilizable. Orifice flowmeters are excluded from consideration. Still, there are enough different types to cause a mental meltdown. Table 2.1 lists most of the in-line methods available to date and the response time constant and rangeability of each. None of the methods listed have enough rangeability to meet the requirements for reagent and air flow measurement for some fed

TABLE 2.1
Typical Flow Measurement Methods

Flow measurement method	Fluid	Accuracy	Rangeability	Time constant
Transmitting rotameter	Liquid or gas	0.5% of rate	11:1	0.2 second
Vortex meter	Liquid	1.0% of f.s.	15:1	2.5 seconds
AC magnetic flowmeter	Liquid	0.25% of rate	20:1	0.01 second
DC magnetic flowmeter & Extended range electronics	Liquid	0.25% of rate	50:1	1.5 seconds
Gyroscopic mass flowmeter	Liquid or gas	0.4% of rate	100:1	0.1 second
Thermal flowmeter	Gas	2.0% of f.s.	20:1	2.0 seconds
Laminar flowmeter	Gas	1.0% of f.s.	50:1	1.0 second

batch fermentors. Thus, flow measurements in these applications must be used as indicators only. The cascade control of pH to flow or DO to flow would do more harm than good. Also, the accuracy figures are the best achievable under controlled test conditions. Field experience shows that the actual errors for industrial applications are two to ten times larger due to variable process conditions and environmental conditions.

The vortex flowmeter infers flow from the rate of pressure or temperature fluctuations caused by vortices shed from a bluff body in the flow path. The meter has a rather large time constant to filter out extraneous perturbations from the signal. Figure 2.1 shows that the repeatability is 0.1%, but the nonlinearity of the meter coefficient (deviation of the coefficient from the horizontal straight line) is 1.0%. Note that the abscissa has the kinematic viscosity in its denominator. The nonlinearity error could essentially be eliminated by the use of polynomials to calculate the viscosity and density as a function of temperature to obtain the kinematic viscosity and by the use of a polynomial to calculate the meter coefficient as a function of the measured vortex frequency in hertz divided by the kinematic viscosity in centistokes. Pipeline vibration and fouling affect those meters that use thermistors to sense the temperature fluctuations. Most meters on the market now use piezoelectric sensors for sensing the pressure fluctuations in order to eliminate vibration and fouling problems and provide a better signal-to-noise ratio (De Vries, 1982).

The magnetic flowmeter infers flow from the voltage induced from the flow of a conducting liquid through a magnetic field. The flow measurement is affected by gas bubbles and falloff of the

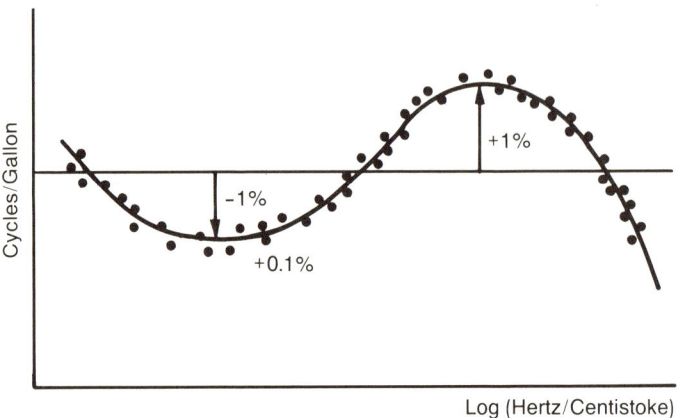

FIGURE 2.1
Vortex Meter Coefficient

electrical conductivity below a threshold value. The fluid must completely fill the meter and be stagnant for zeroing. A series rather than a parallel coil arrangement can create enough heat to damage cells at low flow conditions.

The gyroscopic flowmeter infers flow from the magnetic field changed by the twist in a vibrating U-tube through which the fluid passes. It uses Newton's Second Law of Physics (force = mass · acceleration) to measure the true mass flow rate of the fluid. It is not affected by two-phase flow (solid particles or gas bubbles). However, vibration from the environment affects the reading of older models. As in the magnetic flowmeter, the fluid must completely fill the meter and be stagnant for zeroing.

The thermal flowmeter infers flow from the temperature increase caused by a heating element in the flow path. The flow measurement is affected by the thermal conductivity, the specific heat, and the absolute viscosity of the fluid. The meter's relatively slow response is due to the inherent slowness of temperature sensors in gas streams.

The laminar flowmeter uses Poiseuille's Law for laminar flow to infer flow from a pressure drop that is linear with flow. The main gas stream is subdivided by a matrix into gas streams for the individual capillaries. The laminar flowmeter is capable of measuring extremely low flows (less than 0.0001 scfm) by using fewer and smaller-diameter capillaries. Plugging is usually not a problem, because the gas streams are finely filtered to catch microorganisms. Unfortunately, these meters also respond equally

2.2
Level Measurements

How is a biological reactor similar to a glass of beer? Well, they both give you gas, but, more pertinently, they both have a head of foam. Conductivity probes inserted through the top of a fermentor are used to detect when the head of foam is too large in the vapor space by the decrease in electrical resistance between the probe and the vessel walls or another probe from foam. Needless to say, it doesn't work too well in lined vessels. The probes are used to control the addition of anti-foam agents in an on-off fashion. If the liquid level reaches the probe, the probe will overdose the fermentor with anti-foam agents, which will then suppress the dissolved oxygen level enough to suffocate the cells. This leads to the question of how one goes about measuring the inventory in a fermentor.

For fermentor level measurement, capacitance or admittance probes are not used because the electrical capacitance is too variable; ultrasonic detectors are not used because the foam level is too variable; nuclear detectors are not used because cells don't like radiation; and differential pressure transmitters with diaphragm seals are not used unless pressure-compensated because the pressure is too variable. The most accurate and reliable method is the use of load cells, if properly installed. Load cells measure the weight of a vessel by the amount of strain-induced change in electrical resistance of fine wires bonded to the sides of a support column. Temperature effects are reduced by the use of alloys less sensitive to temperature and suitable compensating resistors in the bridge network. Accuracies of 0.1% of full scale are achievable. The best method of installation uses three or more load cells in tension. Load cells have the added advantage of complete isolation from the fluid. However, the total cost of the installation is relatively large, and the actual liquid level depends upon the degree of aeration and the densities of the broth constituents. Load cells are extensively used for inventory control of the nutrient and substrate storage, and batch charging vessels where the fluid density is well known and the emphasis is more on mass than on level.

2.3
Pressure Measurements

Little needs to be said about pressure measurement except that diaphragm seals are required for sanitary operation. Other than the usual problems experienced with filled capillary systems (i.e., air entrapment in the fill from improper evacuation during assembly and thermal expansion of the fill from changes in the temperature of the process or environment), not much is noteworthy.

2.4
Temperature Measurements

Resistance temperature detectors (RTDs) that measure the temperature-induced change in electrical resistance of a small coil of wire embedded in a ceramic element are preferred over thermocouple sensors because of their better accuracy. Thermocouple sensors have a large bias error from variable alloy compositions and variable sensing junction welds, and a large system error from extension wire and reference junction errors. RTDs can achieve accuracies as good as $0.1°C$ if special care is taken during the specification and installation stages. A three-wire or four-wire system must be specified to compensate for the variable resistance of the lead wires from different lead lengths and temperatures. The temperature span must be narrowed to reduce the nonlinearity error of the sensor resistance versus temperature relationship. The proper resistance versus temperature table must be used for calibration (there are many tables and much confusion as to identification). The ceramic insulator must have a resistance large enough for the given element diameter and length to prevent electrical shunting. The thermowell insertion length must be long enough to reduce thermal shunting. Thermal shunting involves the conduction of heat between the thermowell tip and its vessel connection. The result is a thermowell tip temperature that is different from the true process temperature. Equation 2.1 provides an estimate of the immersion length required for different thermowell constructions and fluid conditions (Richmond, 1980). The immersion length should not be increased much beyond the required length because vortex-induced vibrations can cause mechanical fatigue and eventual failure of the thermowell if it is too long.

$$L = \frac{7}{\sqrt{\dfrac{D}{3 \cdot (D^2 - d^2)} \cdot \dfrac{U}{K}}} \qquad (2.1)$$

where

D = outside diameter of thermowell (inches)
d = inside diameter of thermowell (inches)
U = overall heat transfer coefficient (Btu/hr/sq ft/°F)
K = thermal conductivity of well material (Btu/hr/sq ft/°F/ft)
L = effective immersion length of thermowell (inches)

All of the above discussion concerns steady-state errors. There is also a dynamic error. For a ramping vessel temperature, the measured temperature will lag the true temperature by the total of the time constants for the thermowell assembly. For example, if the vessel temperature increases at the rate of two degrees per minute and the thermowell lag is 45 seconds, the measured temperature will be 1.5 degrees less than the true temperature, as shown by Figure 2.2. What causes most of the lag? Contrary to popular opinion, it is not the mass of the thermowell assembly.

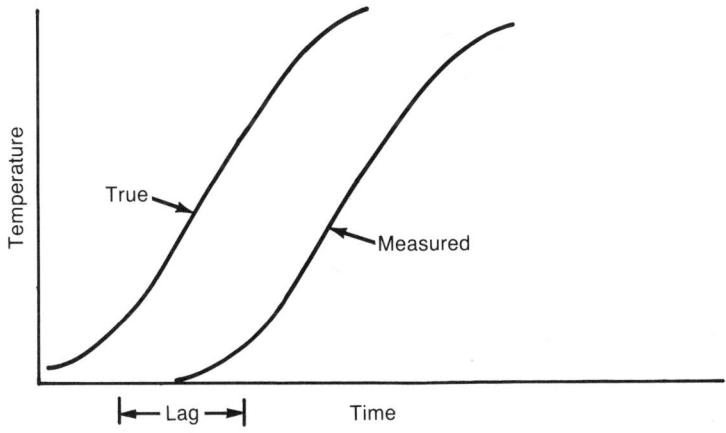

FIGURE 2.2
Dynamic Temperature Error

The culprit is the small annular clearance between the sheathed sensing element and the internal thermowell walls. The air gap in a typical spring-loaded thermowell assembly acts as a great thermal insulator. Table 2.2 shows how the narrowing of this annular clearance and the use of a thermally conductive fill, such as oil, reduce the total of the thermowell time constants (Buckley, 1979). The omission of the thermowell greatly speeds up the response. Table 2.3 shows that a bare sensor element has a single time constant of just a few seconds.

TABLE 2.2
Typical Time Constants for Thermocouples in Thermowells

Fluid type	Fluid velocity (fps)	Annular clearance (inches)	Annular fill	Time constants (seconds)	Total lag (seconds)
Gas	5	0.04	Air	107 & 49	156
Gas	50	0.04	Air	93 & 14	107
Gas	150	0.04	Air	92 & 8	100
Gas	150	0.04	Oil	22 & 7	29
Gas	150	0.005	Air	17 & 8	25
Liquid	0.1	0.01	Air	32 & 10	42
Liquid	1	0.01	Air	26 & 4	30
Liquid	10	0.01	Air	25 & 2	27
Liquid	10	0.01	Oil	7 & 2	9
Liquid	10	0.005	Air	4 & 1	5

TABLE 2.3
Typical Time Constants for Bare Temperature Elements

Bare element type	Fluid type	Fluid velocity (fps)	Time constant (seconds)
Thermocouples			
1/8 inch sheathed & grounded	Liquid	1-3	0.3
1/4 inch sheathed & grounded	Liquid	1-3	1.7
1/4 inch sheathed & insulated	Liquid	1-3	4.5
1/4 inch sheathed & exposed loop	Liquid	1-3	0.1
Resistance temperature detectors			
1/8 inch	Liquid	1-3	1.2
1/4 inch	Liquid	1-3	5.5
1/4 inch dual element	Liquid	1-3	8.0

CHAPTER 3

REDOX and pH Measurements

REDOX measurements are potentially useful wherever oxidation-reduction reactions occur, and pH measurements are potentially useful wherever acid-base reactions occur. REDOX measurements are found mostly in the reaction-conversion area, whereas pH measurements are found in all areas of biochemical processing. Both REDOX and pH measurements can use the same type of reference electrode. There are also many analogous concepts and characteristics between REDOX and pH, some of which will be explored in the following sections.

3.1 REDOX Measurements

While REDOX sounds like some laxative you might buy at your corner drugstore, it is actually the acronym for reduction-oxidation. The REDOX measurement is also commonly called an oxidation-reduction potential (ORP) measurement. As mentioned,

REDOX can be useful in biochemical processes where oxidation and reduction reactions occur. An oxidation reaction involves the loss of electrons by an ion or molecule. Since these electrons must have somewhere to go, an oxidation reaction is always accompanied by a reduction reaction in which an ion or molecule gains electrons. The REDOX potential provides a measure of the degree to which this will occur. The REDOX potential can be viewed as a measure of the electron activity, although this concept has abstract value only, since free electrons do not exist in solutions. A *pe* value can be defined as the negative logarithm of the electron activity to complete the analogy to the hydrogen ion activity and the pH scale. The small "p" designates the mathematical relationship between the charged particle and the variable as a power function, and the "e" designates the charged particle as an electron. Equation 3.1 shows the definition of the *pe* value, and Equation 3.2 shows how the *pe* value can be calculated from potentials associated with the REDOX measurement (Buhler, 1980).

$$pe = -\log(e-) \quad (3.1)$$

$$pe = \frac{E_x + E_r}{E_n} \quad (3.2)$$

where

$e- = $ electron activity (normality)
$E_n = $ Nernst potential of REDOX measurement (mV)
$E_r = $ standard potential of the reference electrode (mV)
$E_x = $ REDOX potential against the reference electrode (mV)
$pe = $ negative logarithm of electron activity (*pe* units)

The REDOX potential depends upon the ratio of the concentration of any reduced species to any oxidized species. It is affected by all species involved in oxidation-reduction reactions. Furthermore, if the reaction involves hydrogen ions, the REDOX potential varies by 8/5ths of the Nernst potential per pH unit, which is 94.7 mV/pH at 25°C. Thus, it is extremely difficult to predict ahead of time how the REDOX potential varies with operating conditions. In practice, test measurements with process samples at various operating conditions must be done to establish a desired set point for REDOX potential.

REDOX AND pH MEASUREMENTS

The test measurements can be used to generate an ORP curve similar to a pH titration curve except that the ORP curve can shift vertically with pH. Figures 3.1(a) and 3.1(b) depict the analogous shapes and abscissa definitions for the curves. ORP curves can have a steep slope near their point of zero excess reduced and oxidized species, just like pH curves that have a steep slope near their point of zero excess hydrogen and hydroxyl ion concentration. Some of the same conclusions about pH loops also apply to ORP loops in terms of the ability to fix the concentration of a

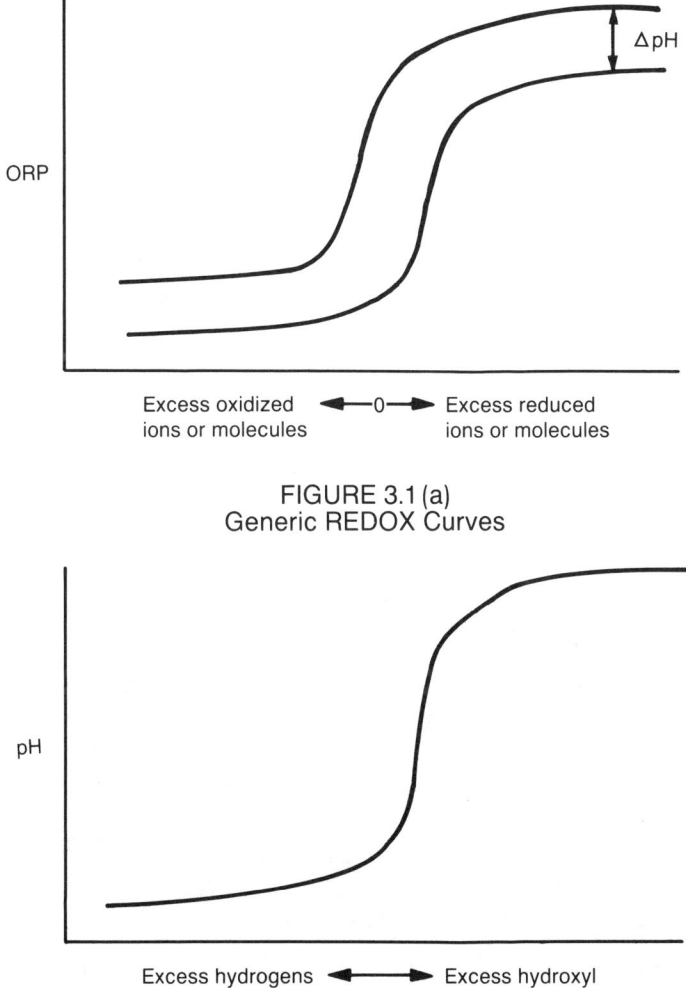

FIGURE 3.1(a)
Generic REDOX Curves

FIGURE 3.1(b)
Generic pH Curve

particular species or its reaction with another species. For a set point on the flat portion of the curve, the process gain is small. A conventional controller can do an excellent job of feedback control. Consequently, the control error is much smaller than the measurement error. The converse situation exists for a set point on the steep portion of the curve. In biochemical processes, the curves for pH are relatively flat due to extensive buffering throughout the control region. In contrast, the curves for REDOX are steeper. Thus, even if the effects of various species and pH could be included to provide a well-defined REDOX curve, the control problem would remain.

To date, ORP measurements are being used as a signature of a fermentor batch rather than for feedback control. In other words, an ORP profile is recorded or logged to distinguish one batch from another. The main advantage gained is more recognition of the change than understanding of the mechanisms involved. Sometimes, the change in ORP can be correlated with a change in an operating condition.

Just as the pH potential is defined by the Nernst equation, so is the REDOX potential. However, the expression for the log involves a ratio of the activities of the oxidized to reduced species, as shown by Equation 3.3a, instead of just the hydrogen ion activity as is the case for pH. Also, the number of electrons in the denominator of the expression for the Nernst potential is greater than one, whereas it is fixed at one for pH. For oxidation-reduction reactions that involve the transfer of just one electron, the Nernst potential at 25° C for REDOX is the same as that for pH (59.12 mV).

$$E_x = E_o + E_n \cdot \log\left(\frac{a_o}{a_r}\right) \qquad (3.3a)$$

$$E_n = 0.1984 \cdot \frac{(T + 273.16)}{n} \qquad (3.3b)$$

where

a_o = activity of oxidized species (normality)
a_r = activity of reduced species (normality)
E_n = Nernst potential (mV)
E_o = standard potential when all activities are unity (mV)
E_x = REDOX potential (mV)
n = number of electrons per oxidation-reduction reactions
T = solution temperature (°C)

A metal measurement electrode is used for REDOX. Only precious metals such as gold and platinum are employed so that the number of metal ions in solution at equilibrium conditions is extremely small. Gold measurement electrodes have a higher potential by about 200 millivolts, but have an exchange current density 12 or more orders of magnitude smaller than platinum electrodes. A higher exchange current density (amperes per centimeter of electrons migrating between the electrode surface and process solution) provides a more correct, more reproducible, and faster responding REDOX measurement. Thus, platinum electrodes are preferred for all applications except for those with strong oxidizing agents that would tend to corrode platinum more than gold (Buhler, 1980).

If the potential of the process solution is above the potential for a platinum-platinum oxide system, the oxygen absorbed on the platinum surface forms a very thin oxide layer. The layer does not affect the sensitivity of the REDOX measurement because it conducts electrons. However, it does affect the speed of response. The oxide layer acts as an oxidation reserve that can retard the decrease of the measured potential when the solution potential has decreased below the potential of the Pt/Pt0 system at the electrode surface. It can also affect the increase of the measured potential until the oxide layer forms when the solution potential increases above the potential of the Pt/Pt0 system. For these reasons, an oxidizing pretreatment should be used for electrodes for applications with high solution potentials, and a reducing treatment should be used for systems with low solution potentials. How high is high and how low is low? Figure 3.2 shows that it depends upon pH. Most biochemical applications would require a reducing pretreatment. This can be accomplished by immersion in a 0.1 mole per liter ferric sulfate solution for a few minutes. Since a rough platinum surface absorbs more oxygen than a smooth one, polished platinum electrodes are preferred (Buhler, 1980).

The millivolt potential error due to a temperature change is much larger for REDOX than for pH in biochemical applications because the pH control band is near neutrality so that the log term multiplier of the Nernst potential is much smaller for pH than REDOX. Interestingly enough, temperature compensation is routinely applied for pH and rarely for REDOX. Fortunately, the temperature control in a fermentor is so tight that temperature compensation is not needed. When calibrating and referencing REDOX measurements, the temperature of the solution should be cited along with the potential.

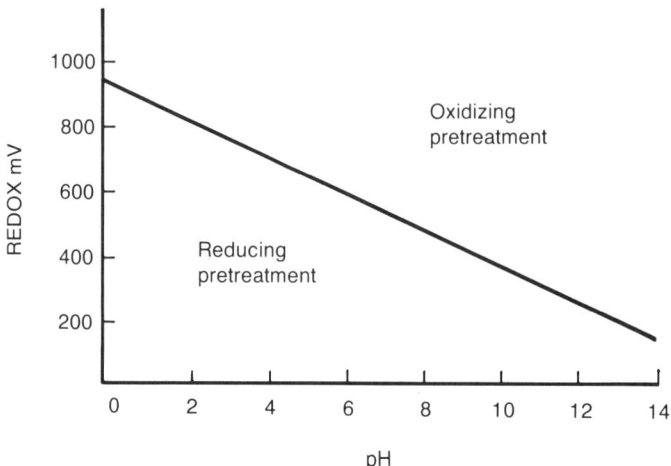

FIGURE 3.2
REDOX Potential of Pt/Pt0 System
(From "Redox Measurements: Principles and Problems," H. Buhler, et al., Ingold publication E-TH 2-1-CH, 1980)

3.2
pH Measurements

pH is the negative logarithm of the hydrogen activity as shown by Equation 3.4. The small "p" designates the mathematical relationship between the ion and the variable as a power function and the "H" designates the ion as hydrogen. Hydrogen ions exist in all fluids that contain water or an acid and can be visualized as a single positive-charged hydrogen nucleus, which is a proton (H). In water streams this hydrogen ion is thought to be actually bound to a water molecule to form a hydronium ion (HO).

$$\text{pH} = -\log(\text{H}) \tag{3.4}$$

where

    H = hydrogen ion activity (normality)
    pH = negative logarithm of hydrogen activity

The actual sensing of pH is accomplished by having a pH-sensitive glass in contact with the internal fill, a pH 7 buffer, and the external sample or stream. The pH-sensitive glass develops potentials per Equations 3.4a and 3.5a, which are Nernst equations, by the hydrogen ion (proton) exchange between hydronium ions in the aqueous solutions, and in the hydrated gel layer of the glass. The protons leave the hydronium ions in the aqueous solution, enter the sites vacated by positive alkali ions such as sodium, and recombine with the hydrogen ions in the hydrated gel layer. The potential developed is proportional to the difference in logarithms of the activity of the hydronium ions in solution and in the gel layer on both sides of the glass membrane. If the gel layers have an equal number of sites for proton exchange, the constants K_{g1} and K_{g2} will be equal. If all the original sodium ions at these sites in the gel are also replaced by protons, the activities a_{g1} and a_{g2} are equal. (Skoog, 1980) If these glass gel constants and activities are equal, Equations 3.4b and 3.5b can be combined to yield Equation 3.6a. By use of the definition of pH per Equation 3.3, the logarithms of hydrogen activity can be converted to pH, which yields Equation 3.6b where the difference in potentials is proportional to the difference in pH. Also, since the internal fill has a hydrogen activity that corresponds to 7 pH, the equation for the potential difference can be simplified to that shown in Equation 3.7. Examination of this equation for the pH measurement electrode yields the following conclusions:

$$E_1 = K_{g1} + 0.1984 \cdot (T+273.16) \cdot \log \frac{a_1}{a_{g1}} \qquad (3.4a)$$

$$E_1 = K_{g1} + 0.1984 \cdot (T+273.16) \cdot [\log(a_1) - \log(a_{g1})] \qquad (3.4b)$$

$$E_2 = K_{g2} + 0.1984 \cdot (T+273.16) \cdot \log \frac{a_2}{a_{g1}} \qquad (3.5a)$$

$$E_2 = K_{g2} + 0.1984 \cdot (T+273.16) \cdot [\log(a_2) - \log(a_{g2})] \qquad (3.5b)$$

If $K_{g1} = K_{g2}$ and $a_{g1} = a_{g2}$, then

$$E_1 - E_2 = 0.1984 \cdot (T+273.16) \cdot [\log(a_1) - \log(a_2)] \qquad (3.6a)$$

$$E_1 - E_2 = 0.1984 \cdot (T+273.16) \cdot (pH_2 - pH_1) \qquad (3.6b)$$

$$E_1 - E_2 = 0.1984 \cdot (T+273.16) \cdot (7 - pH_1) \qquad (3.7)$$

where

a_1 = activity of hydrogen ion in external process fluid (normality)
a_2 = activity of hydrogen ion in internal fill fluid (normality)
a_{g1} = activity of hydrogen ion in outer gel surface layer (normality)
a_{g2} = activity of hydrogen in inner gel surface layer (normality)
E_1 = potential developed at external glass surface (millivolts)
E_2 = potential developed at internal glass surface (millivolts)
K_{g1} = constant for potential for outer gel surface layer (millivolts)
K_{g2} = constant for potential for inner gel surface layer (millivolts)
pH_1 = pH of external solution
pH_2 = pH of internal solution (typically 7 pH)
T = solution temperature (degrees C)

(1) The millivolt output of the electrode decreases as the pH increases.
(2) The millivolt output is zero at 7 pH.
(3) The millivolt output is positive below 7 pH and negative above 7 pH.
(4) The effect of temperature on the millivolt output approaches zero as the pH approaches 7.
(5) At 25°C, the output changes 59.16 millivolts per pH unit.

The accuracy attainable with pH electrodes in a laboratory environment under ideal conditions is impressive. The short-term repeatability for a standard set of electrodes under ideal conditions is +0.01 millivolt. For a solution temperature of 25°C, Equation 3.7 shows that the electrode potential changes by 59.16 mV/pH. Thus, a change of +0.01 millivolt corresponds to a change of less than 0.0002 pH unit.

Besides a repeatability error, there is also drift. The difference in internal and external potential at 7 pH, which is called the asymmetry potential, changes as the electrode glass membrane ages. Under ideal laboratory conditions, the drift is about 0.001

millivolt per day or about 0.00002 pH unit per day (Meriman, 1984).

The accuracy in industrial applications is not as impressive. There are many possible bias, span, and nonlinearity errors from many sources. Even under the best of industrial conditions, pH electrode potential measurements are not more accurate than about +1 millivolt or +0.02 pH (Shinskey, 1973). In critical applications where multiple electrodes are installed, the pH readings are usually not within 0.1 pH of each other. Even though the electrodes are at the same point in the process, just slight differences in electrode orientation and location cause offsets from stray potentials created from concentration and electrical gradients in the stream.

The one error people focus on is the one due to temperature changes, not because it is the most important one, but because it is the one most identifiable. The effort spent on correcting for temperature is almost comical in that no effort is spent on even checking for errors that are an order of magnitude larger. While it is true that the Nernst potential in Equation 3.7 will change with temperature, it is with respect to an absolute temperature and the deviation from 7 pH. At 6.5 pH, an increase in temperature from 25 to 30°C will cause a 0.5-millivolt or 0.01-pH unit error. The widespread use of unnecessary automatic temperature compensators is not a problem *per se* because their complexity, cost, and failure rate are small. It is important to remember that these compensators correct for the change in millivolts developed per pH unit by the electrode and not for the change in pH from the change in dissociation constants of the acids and bases in the process.

An automatic temperature compensator consists of a thermistor immersed in the process with the electrodes, whose resistance changes with temperature to alter the pH amplifier gain in a direction equal but opposite to the change in slope. This resistance is located next to the pH meter span resistance in the feedback path of the field effect transistor (FET) amplifier as shown in Figure 3.3. The compensation is based on the isopotential point (point of zero potential and zero temperature effect) being at 7 pH. In industrial applications, the isopotential point does not stay fixed at 7 pH. Measurement electrode fracture, abrasion, dehydration, etching, or contamination will cause a horizontal shift of the isopotential point as shown in Figure 3.4. Reference electrode contamination and coating, high solution resistance and changes in composition, and high measurement electrode resistance will

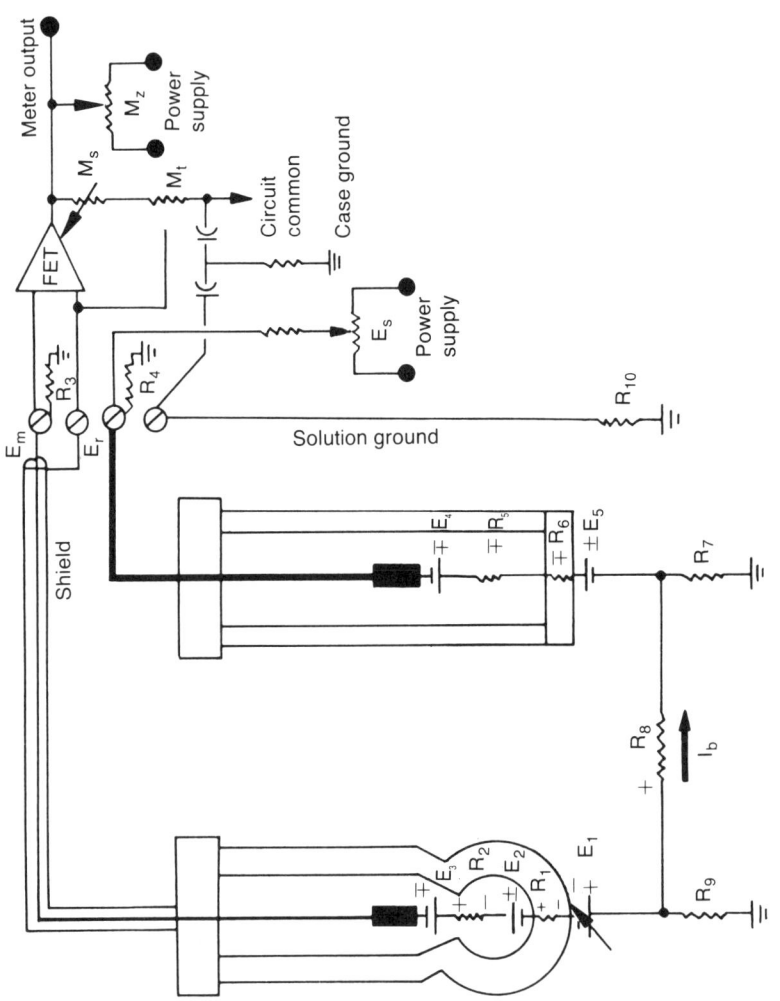

FIGURE 3.3
Equivalent Electrical Circuit for pH Measurement System

REDOX AND pH MEASUREMENTS   (49)

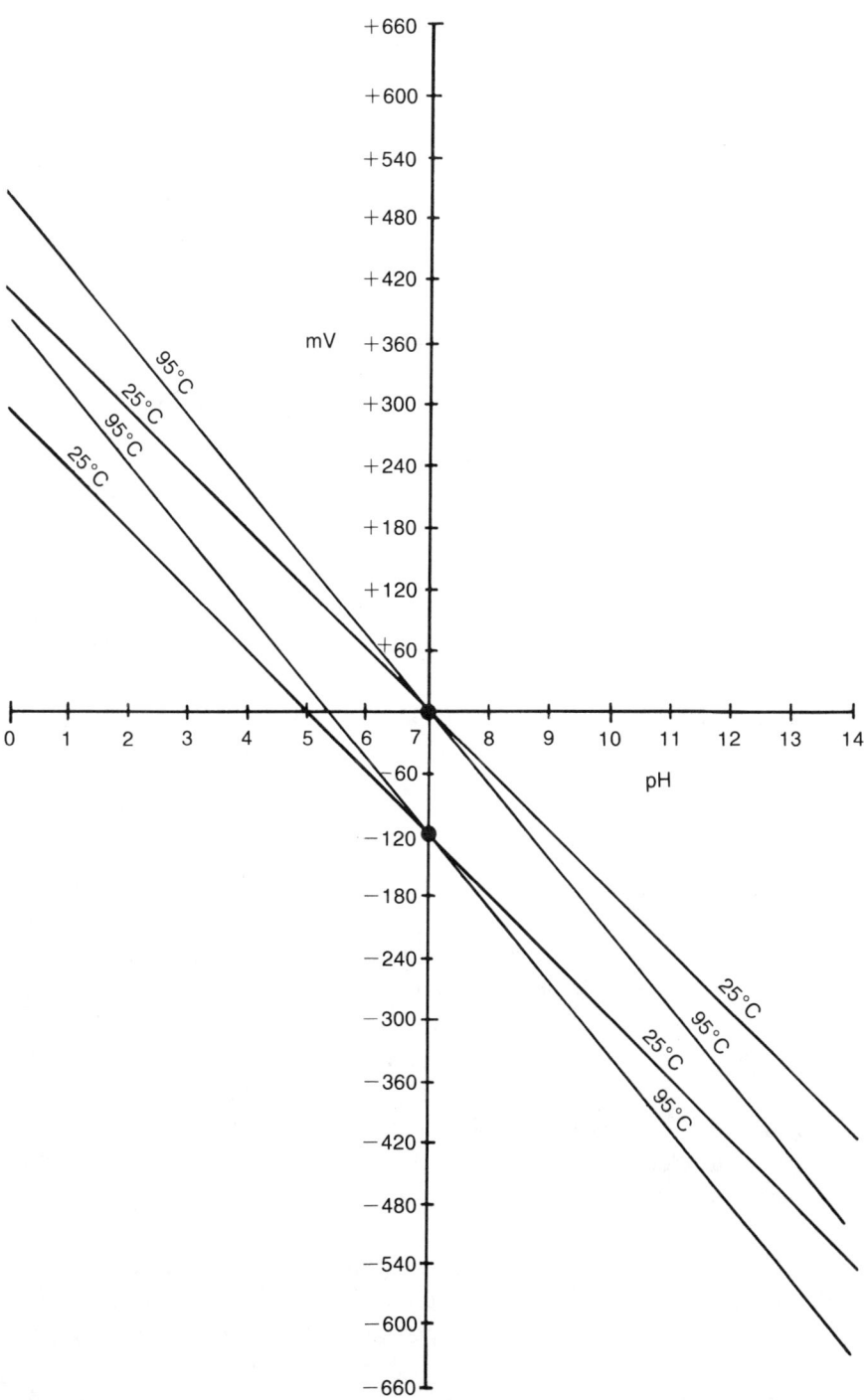

FIGURE 3.4
Horizontal Shift of Isopotential Point

cause a vertical shift in the isopotential point, as shown in Figure 3.5.

If automatic temperature compensation is used, a manual dial is provided to accomplish the same resistance change manually as was done automatically by the thermistor. The main problem with manual temperature adjustment is that the dial reading resolution is poor and the dial location is too accessible. Improper or inadvertent movement of the dial is likely to create more errors than if the temperature effect were ignored.

The number and names of the calibration adjustments vary from one manufacturer to another. The trend is toward consolidation of the adjustments. The older Beckman® pH electronics had the most with five. The meter span adjustment corrected for changes in the millivolt per pH unit slope, the meter zero adjustment corrected for the vertical shifts of the isopotential point, the standardization adjustment corrected for horizontal shifts of the isopotential point, the output span adjustment facilitated reduced spans for better transducer and recorder accuracy, and an output zero adjustment provided elevated zeros (e.g., 2 → 10 pH instead of 0 → 14 pH). Most new pH electronic packages have only two adjustments, typically called span and zero or offset. While simpler to use, much flexibility is lost in correcting specifically for different measurement problems. The general attitude is "why bother" since few users understand what the problems are, let alone how to correct for them, and severe problems should be corrected by electrode replacement rather than electronic recalibration.

Even though pH instruments don't have the ability to correct for many of the pH measurement errors, it is still worthwhile to know how these errors originate so that the hardware, installation, and operating conditions can be changed where possible to reduce errors at their sources. The rest of this section will discuss the nature of these errors. Table 3.1 summarizes the source of the error, the electrical symptom, the response symptom, and the effect on the pH versus millivolt line shown in Figures 3.4 and 3.5.

The potential of interest in pH measurement is the difference between the potential developed at the outer and inner glass surfaces of the measurement electrode as defined by Equation 3.7. Any other potential represents an error. Figure 3.3 shows the location of each potential and Equation 3.8a shows that the effects of these potentials are additive. Whereas changes in the parameters in Equations 3.4 and 3.5 result in either horizontal shifts of the

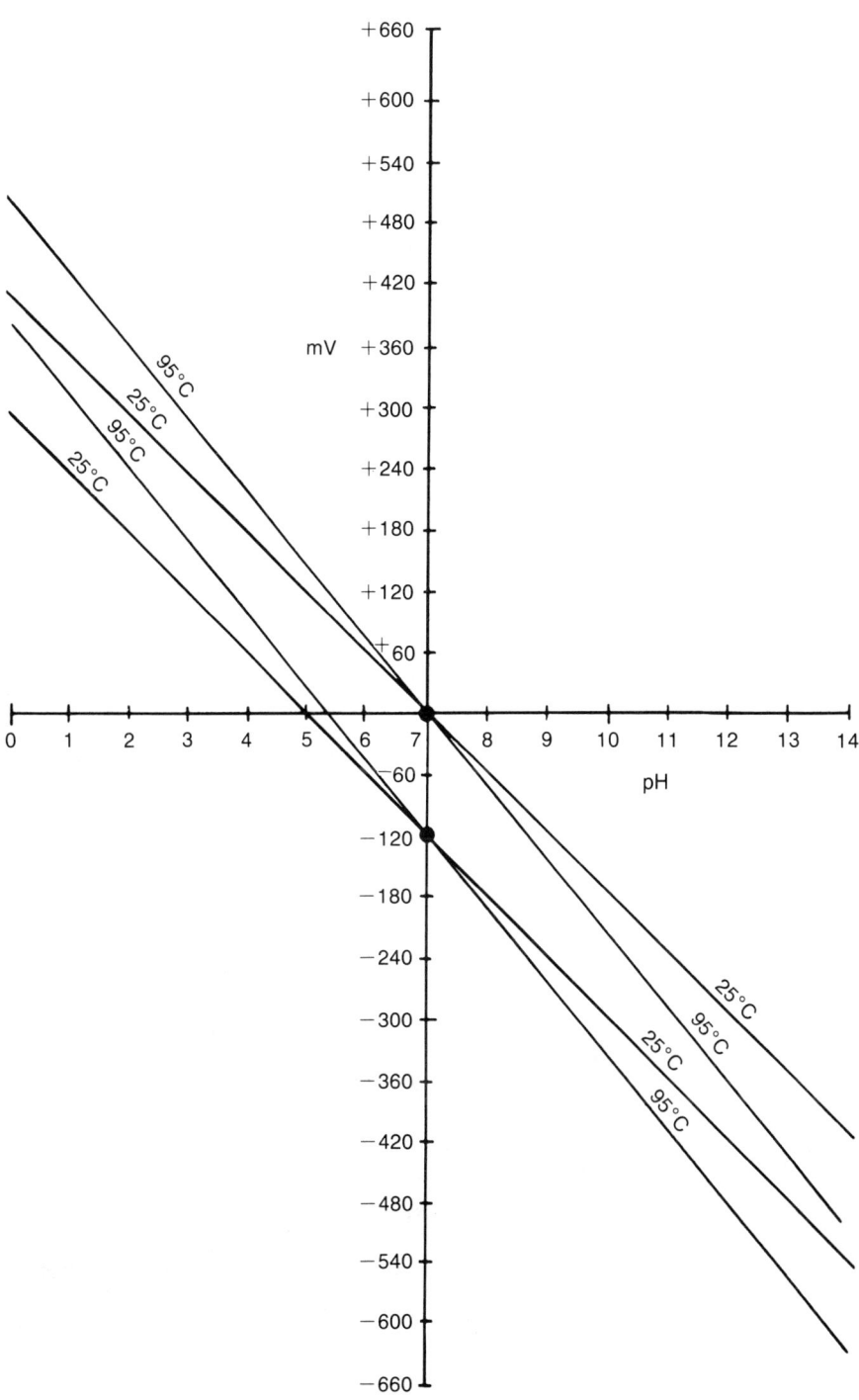

FIGURE 3.5
Vertical Shift of Isopotential Point

TABLE 3.1 Part 1
pH Measurement Errors

Source of error	Electrical symptom	Response symptom	Effect on line in Figure 4.3
Measurement electrode			
Bulb broken	$E_1 = E_2$ & $R_1 \downarrow$	No response (4 to 6 pH)	Horizontal line
Fill contamination	$E_1 = E_2$ & $R_1 \downarrow$	No response (4 to 6 pH)	Horizontal line
Bulb abrasion	$\lvert \Delta E_1/\Delta \text{pH} \rvert \downarrow$ & $E_1 \downarrow$	Slow, erratic, shortened span, & upscale pH	Slope magnitude decreased & isopotential point shifted left
Bulb dehydration	$E_1 \downarrow$, $R_1 \uparrow$ & $\lvert \Delta E_1/\Delta \text{pH} \rvert \downarrow$	Slow, erratic, shortened span, & upscale pH	Isopotential point shifted left & down & slope magnitude less
Bulb etching	$\lvert \Delta E_1/\Delta \text{pH} \rvert \downarrow$ & $E_1 \downarrow$	Slow, erratic, shortened span, & upscale pH	Slope magnitude decreased & isopotential point shifted left
Partial bulb coating	$\lvert \Delta E_1/\Delta \text{pH} \rvert \downarrow$	Very slow	Slope decreased
Complete bulb coating	E_1 fixed	No response	Horizontal line
Low temperature	$R_1 \uparrow$	pH increases as temperature decreases	Isopotential point shifted down
Reference electrode			
Bulb broken	$E_4 \downarrow$ or $\uparrow$	Drift upscale or downscale	Isopotential point shifted down or up
Fill contamination	$E_4 \downarrow$ or $\uparrow$	Drift upscale or downscale	Isopotential point shifted down or up
Partial bulb coating	$E_5 \downarrow$ & $R_6 \uparrow$	Drift upscale (typically)	Isopotential point shifted down
Complete bulb coating	Open circuit	Offscale up or down, depending on meter type	Horizontal line
Thermistor			
Open circuit	$\lvert \Delta E_1/\Delta \text{pH} \rvert \downarrow$	Shortened span	Slope magnitude decreased
Shortened circuit	$\lvert \Delta E_1/\Delta \text{pH} \rvert \uparrow$	Lengthened span	Slope magnitude increased

TABLE 3.1 Part 2
pH Measurement Errors

Source of error	Electrical symptom	Response symptom	Effect on line in Figure 4.3
Solution			
Acidic solvent (no water)	$R_8 \uparrow E_5 \uparrow E_1 \uparrow$	pH offscale downwards	Isopotential point shifted down & increased proton activity
Basic solvent (no water)	$R_8 \uparrow E_5 \downarrow E_1 \downarrow$	pH offscale upwards	Isopotential point shifted down & decreased proton activity
Alcohol solvent (no water)	$R_8 \uparrow$ & $E_5 \downarrow$	pH upscale	Isopotential point shifted down
Hydrocarbon solvent (no water)	$R_8 \uparrow$ & $E_5 \downarrow$	pH upscale & decreased lower & increased upper limits	Isopotential point shifted down & increased line length
Pure water	$R_8 \uparrow$	pH upscale & erratic	Isopotential point shifted down
Composition changes	$E_5 \downarrow$ or $\uparrow$	Drift upscale or downscale	Isopotential point shifted down or up
Gas bubbles	$E_1 \downarrow$ & $R_8 \uparrow$	Upscale noise	Isopotential point shifted left & down randomly
Low pH (acid error)	$|\Delta E_1/\Delta pH| \downarrow$	Shortened span at low pH end	Bends over at low pH end
High pH (alkalinity or sodium ion error)	$|\Delta E_1/\Delta pH| \downarrow$	Shortened span at high pH end	Bends over at high pH end
Low temperature	$|\Delta E_1/\Delta pH| \downarrow$	Shortened span	Slope magnitude decreased
High temperature	$|\Delta E_1/\Delta pH| \uparrow$	Lengthened span	Slope magnitude increased
Terminals			
Short form M to R	$E_i = 0$	Fixed at 7 pH	Horizontal line on abscissa
Broken electrode wire	$E_i = 0$	Fixed at 7 pH	Horizontal line on abscissa
Short form M to ground	$E_i = 0$ & $R_3 = 0$	Fixed at 7 pH	Horizontal line on abscissa
Moisture on M	$R_3 \downarrow$	Stays near 7 pH	Slope magnitude decreased
Moisture on R	$R_4 \downarrow$	Upscale pH	Isopotential point shifts downscale

isopotential point or changes in the millivolt per pH unit slope, all extraneous potentials in Equation 3.8a result in a vertical shift of the isopotential point. The relationship of the Beckman series of calibration adjustments to these potentials is shown in Equations 3.8a, 3.8b, and 3.8c.

$$E_1 = E_1 - E_2 - E_3 + E_4 + E_5 - I_i \cdot (R_1 + R_2 + R_5 + R_6 + R_8) + E_s \tag{3.8a}$$

$$E_o = M_1 \cdot M_s \cdot E_1 + M_z \tag{3.8b}$$

$$I_o = O_s \cdot E_o + O_z \tag{3.8c}$$

where

E_1 = potential developed at external glass surface (millivolts)
E_2 = potential developed at internal glass surface (millivolts)
E_3 = half-cell potential of the measurement electrode (millivolts)
E_4 = half-cell potential of the reference electrode (millivolts)
E_5 = liquid junction potential of the reference electrode (millivolts)
E_i = transmitter input voltage (millivolts)
E_o = meter output voltage (millivolts)
E_s = electrode standardization potential (millivolts)
M_z = meter zero adjustment (bias)
M_s = meter span adjustment (gain)
M_1 = meter temperature compensation adjustment (gain)
O_z = transmitter output zero adjustment (bias)
O_s = transmitter output span adjustment (gain)
I_i = input leakage current of the meter amplifier (milliamps)
I_o = transmitter current output (milliamps)
R_1 = measurement electrode glass resistance (ohms)
R_2 = measurement electrode internal fill resistance (ohms)
R_5 = reference electrode internal fill resistance (ohms)
R_6 = reference electrode liquid junction resistance (ohms)
R_8 = solution resistance between measurement and reference electrodes (ohms)

The measurement and reference electrode half-cell potentials in Equation 3.8a, which are due to an electromechanical reaction between the internal electrodes and fill, are of opposite sign and should ideally be equal so that their sum is zero. However, the half-cell potentials depend upon the electrode type, the internal fill concentration, and the electrode temperature. If the electrode type and fill are identical, then the change in half-cell potential with temperature will cancel out unless a temperature gradient exists between the reference and measurement electrode locations. Contamination of the reference electrode fill will cause a shift of the isopotential point.

Spurious liquid junction potential is located at the tip of the reference electrode. If the ions in the internal fill and the external solution have different mobilities (speeds attained from an electrical force), ion concentration and charge gradients will develop. For instance, the hydrogen ions have a much greater mobility than the chloride ions of a concentrated hydrochloric acid solution. Since ions diffuse to regions of low concentration, the hydrogen and chloride ions will both diffuse to the internal reference electrode fill. If the two ions had identical mobilities, the same number of positive hydrogen ion charges and negative chloride ion charges would arrive at the electrode tip at any given time. However, the hydrogen ion is faster, so that more positive charges arrive and accumulate at the tip. The charge difference between the tip and solution increases until it is large enough to suppress hydrogen ion movement in excess of chloride ion movement to the tip and establish an equilibrium. The isopotential point is shifted up and the pH measurement goes downscale. Table 3.2 shows that the liquid potential size is greatest for concentrated strong acids and that the liquid potential sign is positive for acids and weak bases and negative for concentrated bases.

TABLE 3.2
Liquid Junction Potentials

External Fluid Type	Concentration (Molarity)	Potential (mV)
Hydrochloric acid	1.0	+14.1
Hydrochloric acid	0.1	+ 4.6
Hydrochloric acid	0.01	+ 3.0
Potassium chloride	0.1	+ 1.8
Sodium hydroxide	1.0	− 8.6
Sodium hydroxide	0.1	− 0.4
Sodium hydroxide	0.01	+ 2.3
Potassium hydroxide	1.0	− 6.9
Potassium hydroxide	0.1	− 0.1

Chunks of solids impinging on the electrode, chemical attack of the glass by hydrofluoric acids, or just old age can lead to electrode breakage. The measurement electrode gel layer wears away and penetrates deeper into the glass with time. Even under the best of conditions, the useful life of a measurement electrode rarely exceeds 9 months. If the measurement electrode glass bulb breaks, process fluid will be in contact with both the inside and outside pH-sensitive glass layers. The hydrogen activity, and hence the potential, will be about equal so that the differences in potentials is nearly zero, which corresponds to a constant 7 pH reading.

However, the measurement electrode glass resistance is by-passed (paralleled) by a much smaller process fluid resistance so that R_1 decreases and the measured potential increases per Equation 3.8a. The increase in measured potential corresponds to a decrease in pH reading per Equation 3.7. The result is a shift of the isopotential point vertically upward and a counterclockwise rotation of the millivolt versus pH line to a horizontal position, as shown in Figure 3.6. To determine what the pH reading would be for this position of the line, the original position of the line used in the calibration must be known. For example, if the calibration is represented by the original line position shown in Figure 3.6, the horizontal line in Figure 3.6 at 120 millivolts would give an indication of 5 pH. Field experience shows that a broken measurement electrode causes a constant pH reading between 4 and 6 pH (Meriman, 1983). If the reference electrode breaks, its internal fill becomes contaminated with process fluid. The reference electrode half-cell potential will change and shift the isopotential point vertically upward or downward, and the pH span will accordingly be biased in the opposite direction. Unlike the measurement electrode, the reference electrode can be contaminated without breakage due to its liquid junction. Process pressure, particularly pulsating pressure, can cause contamination that is intolerable for most biochemical applications where measurement accuracy, batch isolation, and microorganism containment must be maintained.

Slurries can cause glass bulb abrasion; nonaqueous fluids or air pockets can cause dehydration; and hydrofluoric acid can cause etching. The response symptoms for glass bulb abrasion, dehydration, and etching are the same in Table 3.1. In each case, the outer glass is less sensitive to the hydrogen activity; or, in other words, a given activity will be measured as a lower activity, which corresponds to a higher pH. The response span is shortened and

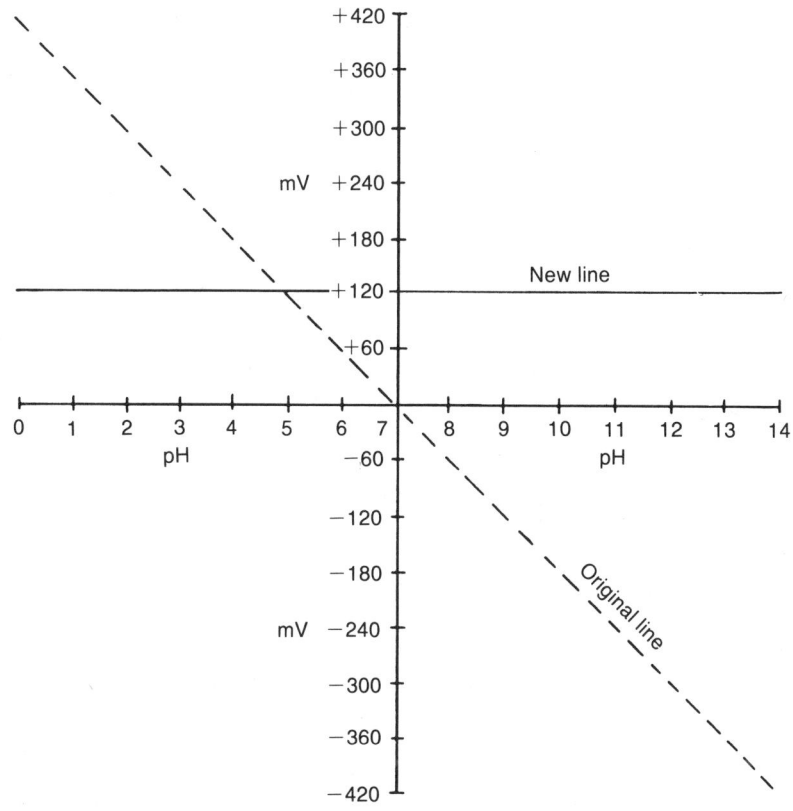

FIGURE 3.6
Measurement electrode breakage results in a constant millivoltage corresponding to 5 to 6 pH reading

offset upscale. If the transmitter is zeroed when the electrodes are immersed in a low pH buffer, the pH reading will be significantly low when the electrodes are immersed in a high pH buffer. The outer glass voltage for a given activity and the change in voltage for a change in activity will be less. The isopotential point is shifted left and the millivolt versus pH line slope is decreased by rotation of the line counterclockwise toward the horizontal position. The increase in glass bulb resistance due to dehydration will cause the isopotential point to shift vertically downward, which aggravates the upscale shift of the pH span. Also, the response may become noticeably slower and erratic due to the reduced and irregular area for proton exchange between the process fluid and the outer gel layer.

Slime, oil, gums, polymers, and particles can cause coating of the measurement and reference electrodes. A partial coating of the measurement electrode forms a layer that slows down the diffusion of the hydrogen ions from the process fluid to the glass surface. The result is a very slow measurement response. The measurement time constant can increase from a few seconds to several minutes. The control loop period will dramatically increase. For a mixer turnover time of one minute and a measurement time constant that increases from two seconds to nine minutes, the loop period will increase from about four minutes to about sixteen minutes (McMillan, 1983). A complete coating of the measurement electrode forms a barrier that stops all diffusion of the hydrogen ions from the process fluid to the glass surface. The result is a constant pH reading whose value depends upon the hydrogen activity of the coating. Since the hydrogen activity of most coatings is low, a constant upscale pH reading is common. This corresponds to a horizontal millivolt versus pH line at a negative millivolt level. A partial coating of the reference electrode will change the liquid junction potential and the resistance at the tip. Usually, the liquid junction potential decreases and the resistance increases so that the isopotential point is shifted down and the pH span is shifted upscale. A complete nonconductive coating can cause an open circuit between the reference and measurement electrodes. The transmitter output will go offscale up or down, depending on the pH meter manufacturer and the model number. The more typical failure mode is upscale, with the pH meter output (transmitter input) approaching a -1 V dc.

Nonaqueous solutions have no water. Nonaqueous systems are rarely found in biochemical processes. Instead of water, the solvent can be an acid, base, alcohol, or hydrocarbon. An acid solvent acts as a proton donor so that the hydrogen activity is increased and the pH scale is shifted down. For example, the pH scale for acetic acid solvent is -6 to -1 pH, and for formic acid solvent it is -9 to -2 pH. A base solvent acts as a proton acceptor so that the hydrogen activity is decreased and the pH scale is shifted up. For example, the pH scale for an ammonia solvent is 16 to 49 pH. An alcohol solvent acts as both a proton donor and acceptor, as does water, so that changes in proton activity are moderated. The pH scale occupies about the same region as water but may extend upscale or downscale. For example, the pH scale for an ethanol solvent is -4 to $+16$ pH. A hydrocarbon solvent acts as neither proton donor nor acceptor so that the solvent is passive to changes in proton activity. Consequently, the pH scale upper and lower limits are usually spread further apart than those for water. For example, the pH scale for an acetone solvent is -5

to +20 pH. Nearly all nonaqueous solvents have a higher resistance than water, so the isopotential point is shifted downscale and the pH span, upscale. This shift is counteracted by the more positive liquid junction potential for acid solvents but is accentuated by the more negative liquid junction potential for base solvents. It is important to realize that dehydration, with all its attendant symptoms and errors, will also occur for all nonaqueous solutions unless the measurement electrode is periodically rinsed with water to replenish the hydronium ions in the gel layer. The start of the dehydration is marked by a slowing down of the response (an increase in the loop period) and a drift of the measurement upscale. In a control loop, the measurement drift would not be noticeable until the reagent valve position or the reagent pump speed had gone to its extreme. Thus, the output of the pH controller should be monitored for early detection of pH measurement drift.

Pure water has a high solution resistance, as do nonaqueous solutions. The addition of a small concentration of ions changes both the conductivity and pH tremendously. The result is an erratic measurement besides the upscale shift of the pH span. Most pure water streams will absorb enough carbon dioxide upon exposure to air to exhibit a buffering effect below 7 pH.

Changes in solution composition will cause changes in the liquid junction potential of the reference electrode. If an acid concentration increases, the liquid junction potential increases and shifts the isopotential point up and the pH span downwards. If a base concentration increases, the liquid junction potential decreases or becomes more negative and shifts the isopotential point down and the pH span upscale.

Gas bubbles can result from sparging, excessive agitation, gas evolution from a reaction, or a gas reagent. The bubbles cause high resistance and reduced hydrogen activity at the measurement electrode upon impact. The result is an intermittent shift of the isopotential point to the left and down, which creates upscale pH noise. The standardization adjustment is used to center the noise band about the proper pH reading so that the short-term plus and minus excursions about the set point tend to cancel out. The transmitter's output is either filtered or the controller's proportional band is increased (gain is decreased) to prevent the reagent valve or pump from reacting to the noise.

Alkalinity or sodium ion error is caused by alkali ions (such as sodium) penetrating the measurement electrode glass-silicon-oxygen network and creating a potential error at the outer electrode glass surface. The result is an indicated pH less than the true pH. The high end of the millivolt versus pH line bends over toward a horizontal position. The error is greater for glasses with alkali ions of equal or larger radius than those of alkali ions in solution. It is frequently called sodium ion error because sodium ions cause a large error due to the penetrating capability of the relatively small ion radius. Alkalinity error increases significantly with temperature.

The cause of the acid error at the low end of the pH scale is not well understood. The onset of the acid error is marked by a reduction in the thickness of the outer hydrated gel layer of the measurement electrode. The result is an indicated pH that is higher than the true pH. The low end of the millivolt versus pH line bends over toward a horizontal position. The sign of the error is opposite that of the alkalinity error. The error is essentially independent of temperature but increases with time.

A short from the measurement to the reference electrode terminals will cause a zero millivolt input and, thus, a 7 pH reading. In fact, the standby position on some pH meters connects a shorting strap between these terminals to give a 7 pH reading and to prevent polarization of the electrodes during immersion of the electrodes or during application of power to the instrument.

3.3
Electrode Installation

The most popular electrode installation assembly until recently was a flow chamber, whose size and shape could vary. A small sample stream was diverted from the vessel or pipeline to the flow chamber. Strainers, filters, and ultrasonic cleaners were added to reduce coating problems. Frequently, plugging of the strainers became as much a maintenance headache as coating of the electrodes. The ultrasonic cleaner proved useful only for loose particulate coatings. The development of the injector type of electrode assembly shown in Figure 3.7 allowed insertion and removal of the electrodes into pressurized vessels and pipelines. More importantly, this installation method eliminated the sample transportation delay associated with sample streams and facilitated high velocities past the electrodes to decrease the coating rate and increase the speed of response.

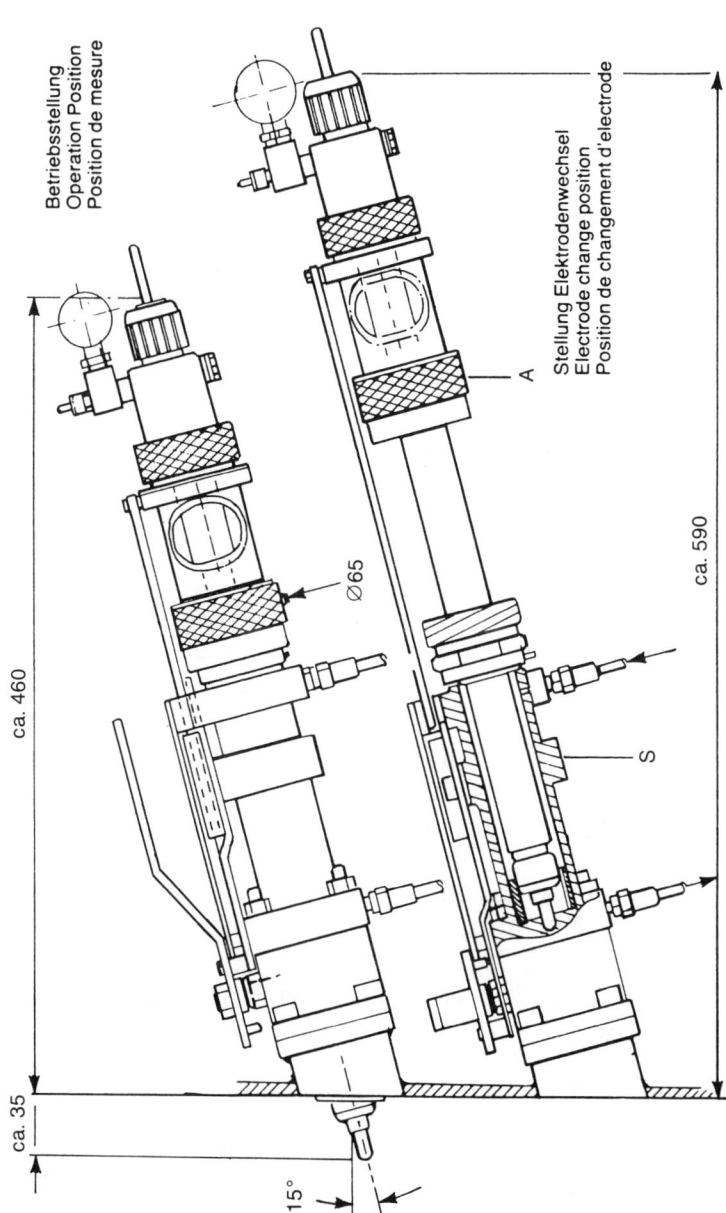

FIGURE 3.7
Insertion Electrode Assembly for Biochemical Applications
(Courtesy of Ingold Ltd.)

Field experience has shown that velocities greater than seven feet per second past the electrodes greatly reduce all types of coating problems. Coating of the reference electrode junction is the most frequent cause of faulty pH measurements in fermentors (Buhler, 1976). The velocity inside even the most highly agitated fermentor does not exceed more than two feet per second except near the impeller tips. The loop fermentor offers an opportunity to reduce coating problems by locating the electrodes in the recirculation line where the velocity is greater than ten feet per second. This location has the added benefits of reducing the measurement noise from gas bubbles and decreasing the measurement electrode response time.

The major source of lag in pH measurement electrode response is the boundary layer of process fluid around the glass bulb that slows down the diffusion and migration of ions to the pH-sensitive glass. For example, the response time to a decrease in pH is about six seconds for a velocity of four feet per second and 48 seconds for a velocity of 0.5 foot per second (the response time stated here and elsewhere in this book is the time required to reach 97.5% of the final value and is roughly equivalent to four times the time constant). Thus, the location of the electrodes to the recirculation line of a loop fermentor causes the net pH measurement response to be faster even though there is a transportation delay from the fermentor to the electrodes.

Pressurization of the reference electrode is rarely used in chemical applications despite the fact that any pressure above atmospheric can force process material back into the reference electrode and contaminate its fill. The requirements for greater measurement accuracy, batch isolation, and microorganism containment in biochemical applications necessitates the pressurization of the reference electrode with sterile nitrogen.

The electrodes must be installed at an angle of 45 degrees or more with respect to the horizontal plane. Otherwise, air bubbles can get trapped in the tip of the electrodes and cause erratic readings.

3.4
Electrode Sterilization

The high temperature of sterilization (typically 121°C for 20 minutes) causes a whole host of problems for REDOX and pH measurement. Table 3.3 summarizes the problems that had to be overcome to have a reliable measurement. The Ingold® series of electrodes have successfully reduced these problems to an acceptable level. For example, after 30 sterilizations, the response time of an Ingold "standard" pH electrode (specifically designed for biochemical applications) increases from 10 seconds to 12 seconds, and the glass resistance increases from 250 megohms to about 350 megohms (Buhler, 1976). These numbers are impressive if one realizes that normal pH electrodes would have been rendered unusable after just a few sterilizations.

TABLE 3.3
Sterilization-induced REDOX and pH Measurement Problems

Measurement component	Sterilization-induced problem
pH measurement electrode	Decrease in mV/pH slope (shortened span)
pH measurement electrode	Increase in response time (sluggish response)
pH measurement electrode	Increase in glass resistance (noise)
pH measurement electrode	Increase in fill AgCl concentration (offset)
Reference electrode	Increase in fill AgCl concentration (offset)
Electrode cables	Permeation of cable jacket by steam (noise)

CHAPTER 4

Dissolved Oxygen and Carbon Dioxide Measurements

Many recombinant DNA microorganisms (affectionately called "bugs") depend upon respiration just like we do. These bugs take in oxygen and give off carbon dioxide. Thus, both oxygen and carbon dioxide are dissolved in the "soup" in which they live. If the soup has too little oxygen or too much carbon dioxide, the bugs suffocate. If the soup has too much oxygen, the bugs get high and out of control. When the oxygen level is not right, the bugs also tend to secrete "stuff" that sours the soup. For example, E coli will secrete acetate ions that must be neutralized by the addition of ammonia to keep the broth pH from dropping. Tight dissolved-oxygen control can eliminate the need for pH control in fermentors.

Dissolved oxygen is measured and controlled in all fermentors where the bugs depend upon oxygen for survival. Now that a sterilizable dissolved-carbon dioxide measurement also exists, dissolved carbon dioxide is starting to be measured on fermentors for diagnostic purposes via trend recordings.

4.1
Dissolved Oxygen Measurement

Dissolved oxygen measurement is amperic; the REDOX and pH measurements described in the last chapter are potentiometric. This classification simply means that the signal developed as a function of the process variable is current for amperic measurements and voltage for potentiometric measurements. The implications of the classification as to sensitivity of the measurement to construction methods and operating conditions is more important than the signals involved. Amperic measurements are affected by electrode area, agitation, and viscosity. Potentiometric measurements are affected by interferences from other ions and absorbable contaminents that block active sites on the sensing surface.

The current actually developed by the dissolved oxygen electrode exponentially increases, reaches a plateau, and then exponentially increases again, as shown in Figure 4.1. The operating point is maintained on the plateau where the current is proportional to the percent of saturation. In order for the operating point of the electrode to be maintained near the center of the plateau, the proper voltage must be available, which for the electrode depicted in Figure 4.1 is about a -0.7 volt. For galvanic electrodes, the voltage is supplied internally by galvanic action, and for polargraphic electrodes, the voltage is supplied externally by a source. Both types of measurements consume the species they measure, but the consumption rate for galvanic measurements is much higher to drive an ammeter directly without amplification. In both cases, oxygen ions are reduced to form hydroxyl ions at the cathode.

The dissolved oxygen electrode actually measures the activity rather than the concentration of oxygen. The activity represents the driving force for the movement of oxygen from the bulk fluid to cathode. Since the oxygen transfer to the cells of biochemical processes depends on the same driving force, the activity rather than the concentration is the process variable of real interest. The concentration is related to the activity via Henry's Law as depicted by Equation 4.1 (the activity is expressed as the partial pressure of oxygen). Henry's coefficient depends upon the temperature and the composition of the fluid. For example, a dissolved oxygen electrode will read the same in a potassium chloride solution and a pure water solution that are in equilibrium with air at the same

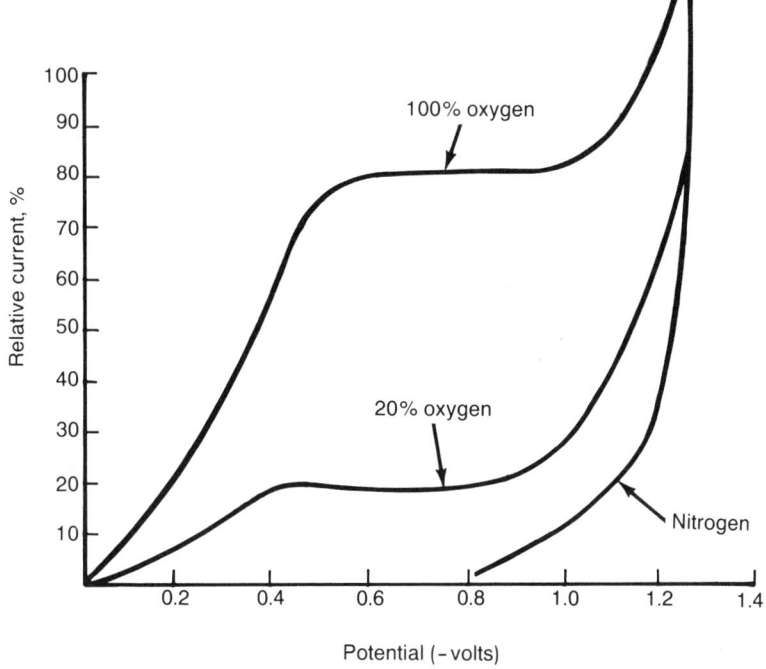

FIGURE 4.1
Typical Current Voltage Curves for DO Electrode

temperature, even though the concentration of the pure water solution is ten times larger. If the solutions are agitated, the readings will also be the same as that for the air. The dissolved oxygen measurement is usually indicated as a percent of saturation, which can be converted to partial pressure in millimeters of mercury by multiplication by the atmospheric pressure (760 mm at sea level).

The current developed by an electrode for a given partial pressure of oxygen can be approximated by Equation 4.2a. On a first principle basis, the current is equal to the driving force (the product of the partial pressure and the cathode surface area on which it acts) divided by the total impedance of the installed sensor. The impedance of greatest effect and variability is that associated with the membrane. Equation 4.2c shows that this impedance is proportional to the membrane thickness and is inversely proportional to the membrane permeability. Equation 4.2d shows the membrane permeability is proportional to the product of the diffusion coefficient and solubility of oxygen in the membrane and an exponential temperature relationship. Equations 4.2a, 4.2c, and 4.2d are combined to give Equation 4.3, which shows that the greatest current is generated by electrodes that have large cathode areas and thin membranes with high oxygen solubilities that are in solutions at high temperatures. Unfortunately, the oxygen consumption is proportional to the current. High oxygen consumption increases the zone of oxygen depletion that extends from the membrane into the solution. Once the zone has developed, its extent is affected by the degree of agitation. Thus, electrodes with high current production also tend to have high flow sensitivity. The galvanic electrode has the highest sensitivity for this very reason. Table 4.1 shows that the reading in stagnant water is about one half that in stirred water. The other electrodes listed show much less sensitivity because an optimum construction was designed to minimize the inaccuracies associated with both current magnitude and oxygen depletion (Krebs, 1972).

$$P_o = H \cdot X_o \tag{4.1}$$

$$I_o = K \cdot \frac{A}{Z_t} \cdot P_o \tag{4.2a}$$

$$Z_t = Z_e + Z_m + Z_f \tag{4.2b}$$

$$Z_m = K \cdot \frac{d}{P_m} \tag{4.2c}$$

$$P_m = K \cdot D \cdot S \cdot e^{-1/T} \tag{4.2d}$$

if $Z_e + Z_f << Z_m$, then

$$I_o = K \cdot \frac{A \cdot D \cdot S \cdot e^{-1/T}}{d} \cdot P_o \tag{4.3}$$

where

A = area of cathode (sq cm)
d = thickness of membrane (micrometers)
D = diffusion coefficient of oxygen in membrane (sq cm/sec)

H = Henry's Coefficient (mm Hg/mole fraction)
I_o = current output of electrode (microamps)
K = proportionality and units conversion factor
P_m = permeability of membrane (moles/sec)
P_o = activity expressed as partial pressure of oxygen (mm Hg)
S = solubility of oxygen in membrane (moles/sq cm)
T = absolute temperature of fluid (°K)
X_o = concentration of dissolved oxygen (mole fraction)
Z_e = mass transfer impedance of electrolyte layer (mm Hg · sq cm/microamp)
Z_m = mass transfer impedance of membrane (mm Hg · sq cm/microamp)
Z_f = mass transfer impedance of process fluid (mm Hg · sq cm/microamp)

TABLE 4.1
Flow Sensitivity of DO Electrodes

Type of electrode	Output in air	Output in stagnant water	Output in stirred water
Polargraphic (blood-gas sample)	100%	99.5%	100%
Polargraphic (fermentation)	100%	98.5%	100%
Galvanic	100%	50.0%	99%

The speed of response also gets faster for thin and highly permeable membranes. However, thin membranes are more susceptible to the occurrence of pin holes, which will lead to electrolyte contamination. Also, thin membranes will not withstand the high pressures during fermentation of the high temperatures during sterilization. Membranes that are very permeable to oxygen are usually very permeable to water as well. In fact, Teflon® membranes are six times more permeable and dimethyl silicone rubber membranes are sixty times more permeable to water than oxygen. This can cause water loss from the electrolyte to the process and eventual drying of the inner electrode fill (Krebs, 1972).

The effect of temperature is not strictly exponential as shown in Equation 4.3 because of various nonidealities. A change of 2 percent in reading for each °C change in temperature can be used as a rule of thumb. The effect can be reduced to about 0.05 percent per degree by the use of a thermistor and a single operational amplifier. Temperature compensators have not been extensively used to date on dissolved oxygen measurements even though the need for them is obviously great (Krebs, 1972).

The basic construction of a dissolved oxygen electrode is shown in Figure 4.2. The electrode typically uses a silver anode, saturated silver chloride electrolyte, a platinum cathode, and a polymeric membrane. In order for such an electrode to withstand the rigors of fermentation applications, a special double layer configuration of the membrane has been developed by Instrument Laboratory, Inc.® The inner layer is a medium permeability Teflon film that is 25 micrometers thick. The outer layer is a high-permeability silicon membrane reinforced by thin steel netting that is 125 micrometers thick. The thicker outer layer increases the operating pressure capability and response time and decreases the

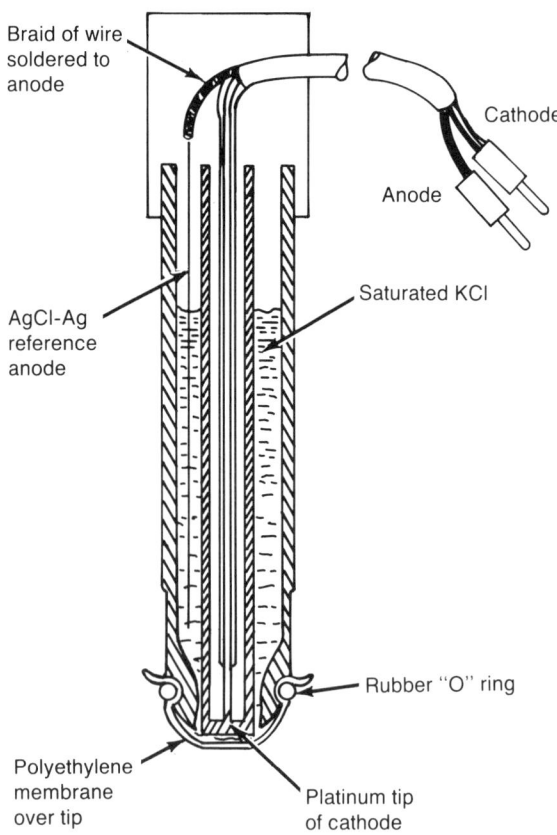

FIGURE 4.2
Construction of Typical Dissolved Oxygen Electrode
(Courtesy of Instrument Laboratory, Inc.)

flow dependence of the electrode. The response time has been measured to be about 50 seconds (Buhler, 1976).

Many recordings of dissolved oxygen show a significant amount of measurement noise. Most of this noise is due to air bubbles momentarily attached themselves to the membrane. Due to dynamic conditions, the oxygen in the air bubble and in the broth are not exactly in equilibrium with each other so that the activity of oxygen in the gas and liquid are not quite equal. Even if the activities were equal, the measured oxygen for the galvanic type of electrode can differ by 50 percent for gas and liquid per Table 4.1. Since the galvanic type is the simpler and less expensive type of electrode, many university laboratories use them for their research. Since most of the publications on biochemical measurement and control are from academia rather than industry, the figures show a large noise band for dissolved oxygen measurement.

The dissolved oxygen electrode is calibrated by the use of nitrogen to set the zero adjustment and air to set the span adjustment. Since the electrode responds very fast and accurately to oxygen changes in the gas phase, it is preferable to remove the probe from the fermentor for a gas calibration rather than bubbling gases through a precharge of substrate for a liquid calibration. Any current during exposure to nitrogen is called a background, zero, offset, or nitrogen current. It is usually caused by an improper polarization voltage or back-diffusion of oxygen. A polarization voltage to the right of the plateau in Figure 4.1 will cause water reduction and an exponential increase in current. Long exposures of a probe to oxygen cause an equilibrium to be reached where oxygen has accumulated in the electrolyte. Upon removal of the oxygen source, oxygen diffuses out through the membrane and back to the cathode, where it is reduced with an accompanied amount of current generation. The background current for a well-designed electrode should not exceed one percent of full scale, which is within the accuracy spec of the measurement.

4.2
Dissolved Carbon Dioxide Measurement

The dissolved oxygen measurement provides information on whether the oxygen supply is available to the cells. The dissolved carbon dioxide measurement provides verification that the cells were actually able to use the oxygen. It confirms that the cells are healthy and happy with their environment.

Like the dissolved oxygen probe, the dissolved carbon dioxide probe measures the partial pressure or activity of a molecule. However, the dissolved carbon dioxide measurement is a potentiometric measurement. Thus, deposits on the membrane or changes in the membrane thickness will primarily affect the response time rather than the final measured value. Deposits from anti-foam agents, which play havoc with a dissolved oxygen measurement, have relatively little effect on a dissolved carbon dioxide measurement.

The dissolved carbon dioxide measurement is really a disguised pH measurement. A pH electrode is immersed in a bicarbonate electrolyte within the probe. Carbon dioxide diffuses through a gas-permeable membrane into the electrolyte until the concentration in the electrolyte is in equilibrium with the partial pressure of carbon dioxide per Henry's Law. The hydrogen ion concentration in the electrolyte is proportional to the carbon dioxide concentration in the electrolyte per Equation 4.4 for the dissociation constant, since the bicarbonate concentration is so large that it is essentially constant. The potential of the pH electrode is proportional to the log of the hydrogen ion activity and, hence, the log of the partial pressure of carbon dioxide in the process fluid per Equation 4.5. To summarize, the hydrogen ion concentration is proportional to the hydrogen ion activity because the activity coefficient is constant in the relatively fixed environment of the electrolyte. The carbon dioxide concentration is proportional to the hydrogen ion concentration due to the high bicarbonate concentration in the electrolyte. The carbon dioxide partial pressure is proportional to the carbon dioxide concentration via Henry's Law. It is important to realize that changes in the pH and the concentration of solutes in the process fluid do not affect the reading unless they change the partial pressure of carbon dioxide (Puhar, 1980).

$$K_a = \frac{|H+| \, |HCO_3-|}{|CO_2|} \tag{4.4}$$

$$E_c = E_o + E_n \cdot K \cdot \log(P_c) \tag{4.5}$$

where

$|H+|$ = hydrogen ion concentration (normality)
$|HCO_3-|$ = bicarbonate ion concentration (normality)
$|CO_2|$ = carbon dioxide concentration (normality)
E_c = dissolved carbon dioxide potential (mV)
E_n = Nernst potential (mV)
E_o = standard potential for unity hydrogen activity (mV)
K = units conversion factor
K_a = dissociation constant for carbonic acid
P_c = partial pressure of carbon dioxide (mm Hg)

Many fermentors operate at a pH above the negative logarithm of the dissociation constant (pK_a) for carbonic acid. At this point the apparent solubility of carbon dioxide is much larger than the real solubility per Equation 4.6. Since the partial pressure of carbon dioxide is proportional to the real dissolved carbon dioxide concentration, the partial pressure and concentration of carbon dioxide in the off-gas will decrease with an increase in pH. Since the pK_a changes with temperature ($pK_a = 6.322$ at 30°C and $pK_a = 6.357$ at 37°C), the effect also increases with temperature, as shown in Figure 4.3. Calculations to date of the carbon dioxide production rate (CPR) and the respiratory quotient (RQ = CPR/OUR where OUR is oxygen uptake rate) have ignored the effect of pH on the amount of carbon dioxide measured in the off-gas. Changes in either the CPR or RQ caused by changes in the solubility of carbon dioxide from pH and temperature changes must be distinguished from changes due to cell metabolism for cell growth models used in optimization and computer control strategies.

$$S' = S \cdot [\text{antilog}\,(\text{pH} - pK_a) + 1] \tag{4.6}$$

where

pH = negative logarithm of hydrogen activity
pK_a = negative logarithm of dissociation constant
S = real solubility of carbon dioxide (moles/cu cm)
S' = apparent solubility of carbon dioxide (moles/cu cm)

Carbon dioxide electrodes have been used in clinical laboratories for many years but were not used in industrial applications

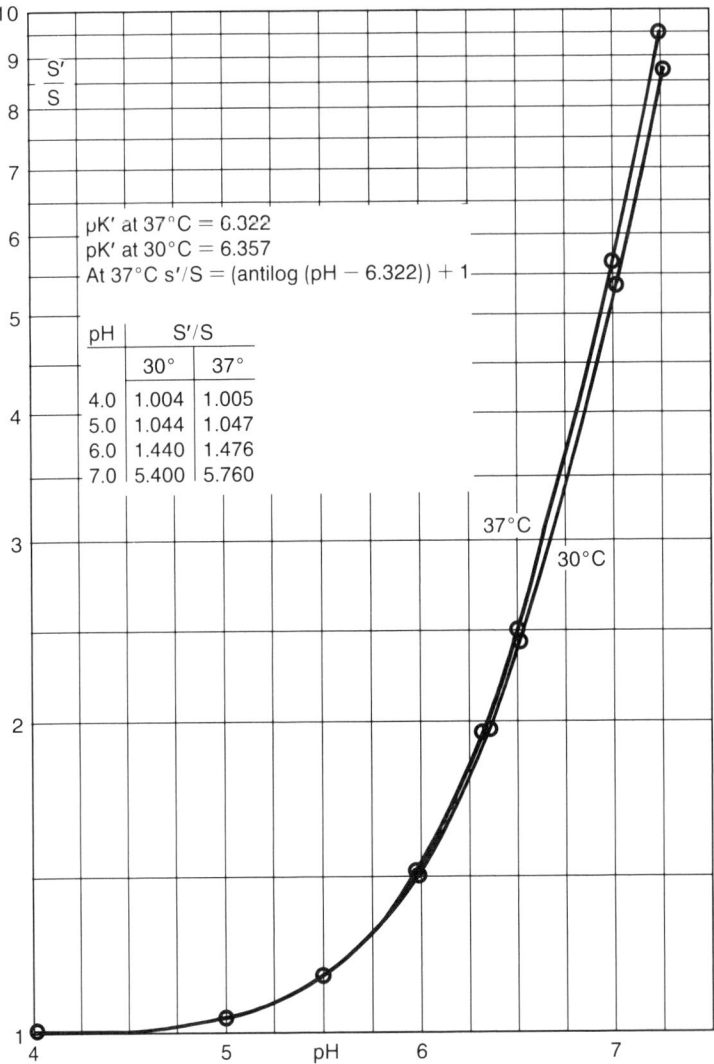

FIGURE 4.3
Effect of pH on Solubility of Carbon Dioxide

until after the development by Ingold, Ltd.® of a probe that is both rugged and maintainable. The probe consists of a sterilizable pH electrode placed inside a stainless steel tube filled with bicarbonate electrolyte, as shown in Figure 4.4. The probe tip has a silicon membrane reinforced by a stainless steel mesh. A nylon net is interposed between the membrane and the pH electrode bulb to insure a constant thickness of the electrolyte film. The pH electrode is easily exchangeable. The probe is calibrated by

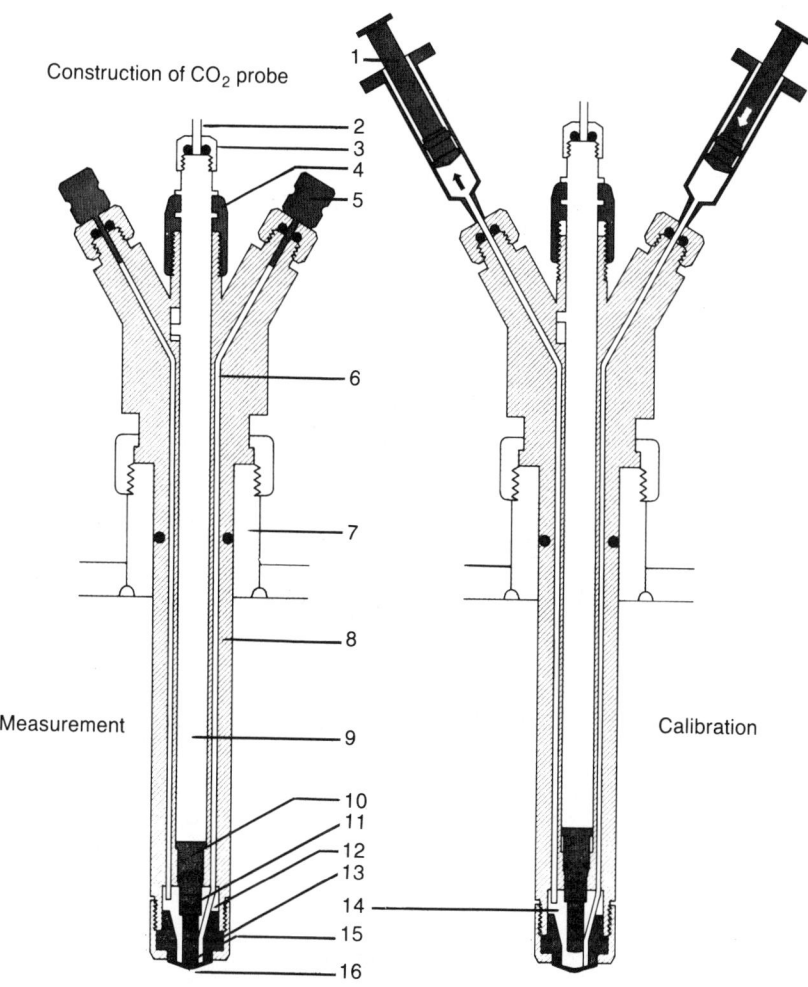

FIGURE 4.4
Construction of an Industrial Dissolved Carbon Dioxide Probe
(Courtesy of Ingold, Ltd.)

slightly withdrawing the pH electrode, withdrawing the electrolyte with the left syringe, and injecting the buffer with the right syringe as shown at the right in Figure 4.4. The isopotential point shift that normally occurs after sterilization can be compensated by an adjustment of the standardization potential in the pH meter electronics. The accuracy of the measurement after such recalibrations is about $+10$ percent of the reading. Dissolved carbon dioxide probes can not be calibrated by exposure of the probe to air, as is done with dissolved oxygen probes, because the carbon dioxide content of air is small and not well-defined and the probe response to such small concentrations is very sluggish. The probe response time for the carbon dioxide levels normally encountered in fermentations is one to two minutes. If the electrolyte film composition becomes altered by the diffusion of water, organic acids, ammonia, or other contaminents, and old electrolyte can be easily replaced by fresh electrolyte by use of the syringes. During a long fermentation, the electrolyte can be replaced and any change in reading used as an indication of the extent of contamination.

CHAPTER 5

Composition Measurements

The intricacy of a biochemical process compared to that of a chemical process is like the intricacy of a snowflake compared to that of a raindrop. The complexity and number of different reacting molecules in a fermentor is staggering. Composition measurements are needed so that biochemical plant operation does not become a magical mystery tour.

To date, nearly all biochemical reactors (i.e., fermentors) are operated without a direct or inferred composition measurement. The pH, dissolved oxygen, and temperature measurements commonly used on fermentors provide information on broth environment rather than on the broth composition. Fermentor product quality relies on the ability to duplicate the feeding pattern and operating conditions arrived at after years of deliberate and accidental trial and error experimentation in the laboratory, pilot plant, and industrial plant. The greatest revolution in industrial biochemical measurement and control will occur when the myriad of new analytical measurement techniques on the horizon are commercialized. While the present emphasis is on reaction/conversion area applications, the impact on separation/purification area product quality is even greater.

5.1
Biosensors

A biosensor consists of an enzyme (a highly selective and ultra-sensitive catalyst) immobilized on a thermistor or ion-selective electrode. The enzyme is chosen based on its ability to promote a reaction of the compound to be measured. When used with a thermistor, the biosensor infers the reactant concentration from the heat of reaction. Temperature changes from any other source cause errors unless compensated for by another thermistor. When used with a ion-selective electrode, the biosensor infers the reactant concentration of interest by measuring the uptake of one of the reactants of liberation of one of the products of the catalyzed reaction.

Glucose is the most common substrate in fermentors. The cell growth rate can be inferred from the glucose consumption and the growth rate controlled by glucose concentration control. Too little glucose during the batch can cause cell starvation and too much glucose at the end of the batch represents a wasted raw material whose cost is the largest part of the production cost. Short-term glucose depletion has been shown to occur during supposedly routine operation. Without a glucose measurement, such depletions are not identified. The result is unresolved variations in product quality or quantity between batches.

Glucose can be measured by the immobilization of the enzyme called glucose oxidase on a dissolved oxygen electrode. The glucose concentration is inferred from the oxygen uptake in the catalyzed reaction of glucose and oxygen to form hydrogen peroxide and gluconic acid per Equation 5.1. Alternately, the gluconic acid liberation could be measured by a pH electrode or the hydrogen peroxide liberation could be measured by a platinum electrode. The life of the probe largely depends upon the method of entrapment of the enzyme. A probe with a soluble enzyme held in place by a dialysis membrane will normally last one week; a probe with physically secured enzyme is normally good for a week; and a probe with a chemically bound enzyme can normally be used for over a year (Guilbault, 1981).

Perhaps the second most discussed biosensor is the urea electrode. The enzyme urease is immobolized on a ion-selective electrode to sense the ammonium ion produced per Equation 5.2 or onto a gas membrane electrode to sense the ammonia gas liberated in the presence of hydroxyl ions per Equation 5.3. The ammonium ion electrode is similar to the pH electrode but with a different composition of the glass and internal fill. Unfortunately, this electrode also responds to the sodium and potassium ions. For the analysis of urea in blood and urine where these ion concentrations are high, the ammonia gas electrode is preferred. It is similar in construction to the carbon dioxide electrode described in Chapter 4. The ammonia gas diffuses through a membrane into an ammonium buffer, and the resulting change in pH is measured by a glass pH electrode.

The biosensors designed to date are not suitable for mounting directly in a fermentor. These sensors are not capable of being steam sterilized in place. Also, they generally require some sample preconditioning. The on-line installation depicted in Figure 5.1 for glucose analysis shows a slipstream from a small fermentor pumped to a system for microfiltration, buffer addition, mixing, and heating. The cells are removed from the sample and recycled back to the fermentor to reduce the loss of product and to prevent coatings on the probe that would reduce the enzyme activity. The temperature and pH of the filtered sample is then regulated at the optimum value for greatest enzyme activity. The optimum pH for the glucose and urea electrodes is 7 pH. In general, the optimum pH for biosensors lies between 3 and 9 pH (Guilbault, 1981).

Reaction at the glucose electrode:

$$\text{glucose} + \text{oxygen} \xrightarrow{\text{glucose oxidase}} \text{hydrogen peroxide} + \text{gluconic acid} \tag{5.1}$$

Reactions at the urea electrode:

$$\text{urea} \xrightarrow{\text{urease}} \text{ammonium ion} + \text{carbonic acid ion} \tag{5.2a}$$

$$\text{ammonium ion} \xrightarrow{\text{hydroxyl ions}} \text{ammonia} \tag{5.2b}$$

More than 40 different biosensors are under development at the time of this writing. The more notable ones and some of their characteristics are described in Table 5.1. The response time listed in this table is the time required to reach 95 percent (three time constants). It should be multiplied by 4/3 to get the 98 percent (four time constants) response time. Physical immobilization was

(80) COMPOSITION MEASUREMENTS

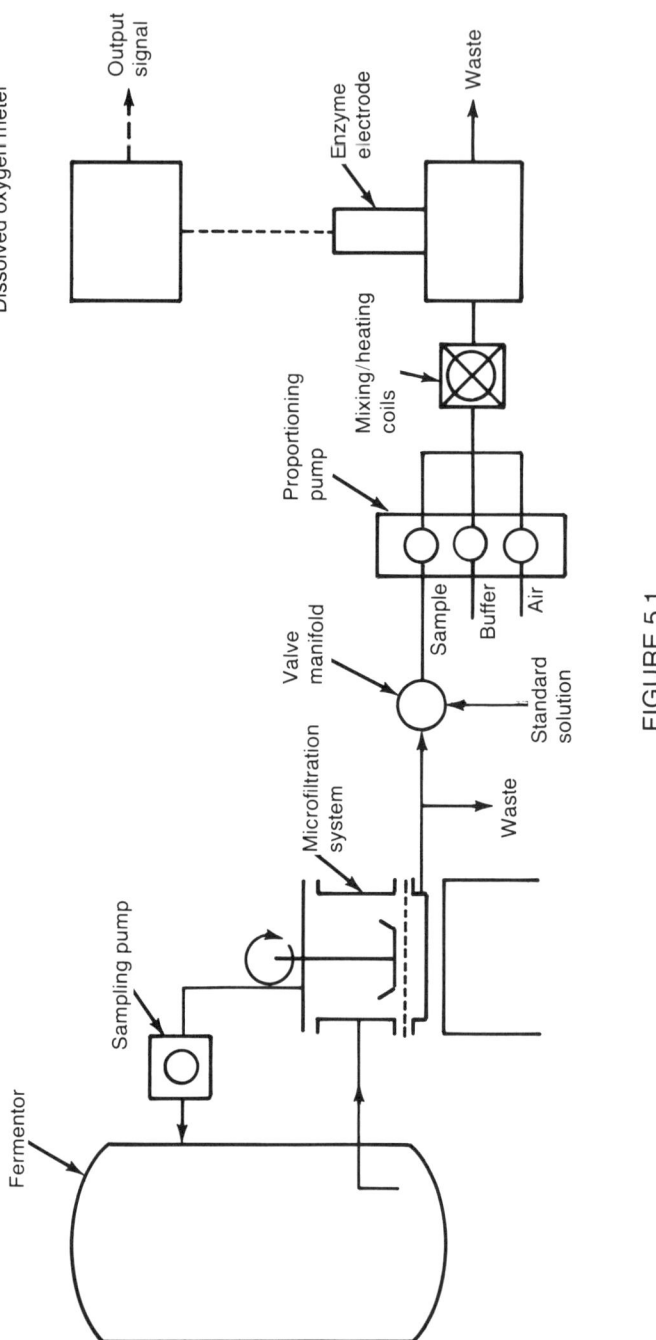

FIGURE 5.1
On-line System for Glucose Analysis
(Source: Chotani, 1982)

TABLE 5.1
Typical Response Characteristics of Enzyme Electrodes
(Source: Guilbault, 1981)

Compound measured	Immobilization method	Useful life	Response time	Range (moles/liter)
L-Amino acids	Chemical	4–6 months	2 min	0.01–0.0001
	Physical	2 weeks	1–2 min	0.01–0.0001
L-Tyrosine	Physical	3 weeks	1–2 min	0.1–0.0001
L-Glutamine	Soluble	2 days	1 min	0.1–0.0001
L-Asparagine	Physical	1 month	1 min	0.01–0.00005
L-Lysine	Chemical	6 months	5 min	0.01–0.00005
L-Methionine	Chemical	6 months	2 min	0.01–0.00001
D-Amino acids	Physical	1 month	1 min	0.01–0.00005
Alcohols	Soluble	1 week	1 min	0.01–0.00005
	Chemical	>4 months	30 sec	0.01–0.00005
Amygdalin	Physical	3 days	2 min	0.01–0.0001
Cholesterol	Chemical	2 months	5 min	0.01–0.0001
Glucose	Soluble	1 week	5–10 min	0.1–0.001
	Physical	3 weeks	2–5 min	0.01–0.0001
	Chemical	>14 months	1 min	0.02–0.001
Penicillin	Soluble	3 weeks	2 min	0.01–0.0001
	Physical	2–3 weeks	0.5–2 min	0.01–0.0001
Phosphate	Chemical	4 months	1 min	0.01–0.0001
Sulfate	Chemical	1 month	1 min	0.01–0.0001
Urea	Physical	2–3 weeks	1–2 min	0.01–0.0001
	Chemical	>4 months	1–2 min	0.01–0.0001
Uric acid	Chemical	>4 months	30 sec	0.01–0.0001

achieved by entrapment with a polyacrylamide gel, and chemical immobilization was accomplished by attachment to glutaraldehyde with albumin, polyacrylic acid, or glass beads (Guilbault, 1981).

5.2
CHEMFETs and ISFETs

Chemically sensitive field effect transistors (CHEMFETs) and ion-sensing field effect transistors (ISFETs) promise to revolutionize the general area of on-line chemical analysis due to their small size (about that of a dime) and low cost (about $100). The result is a throwaway sensor. Instead of worrying about whether a sensor will take repeated sterilizations or be depleted or coated after sustained use, it can be pitched and replaced. The hardware cost is less than the cost of the technician's time spent to repair or calibrate an old sensor. The miniaturization of these devices is possible through the use of microlithography, the technology developed for the manufacture of chips for integrated circuits. Thin metallic and chemically sensitive films are vacuum deposited on silicon substrates. A pattern is formed in the film by applying a light-sensitive organic coating, illuminating areas to make them more soluble, washing away these exposed areas, and etching the film in the areas not covered by the organic coating (Weiss, 1985).

A CHEMFET or ISFET functions like a normal field effect transistor except the field at the gate is not a function of a control voltage but of the ions of the process fluid in contact with the chemically sensitive coating on the gate. The response is close to that predicted by the Nernst equation for many common ions. It has been used to detect the presence of enzymes and antibodies. Figure 5.2 shows the general configuration of a generic CHEMFET or ISFET. The figure does not show the detailed construction of the sensor, which is designed so that only the chemically sensitive gate is in contact with the process. Presently, these sensors are in their infancy. They have not as yet moved from the laboratory development phase to test installations in industrial applications (Weiss, 1985).

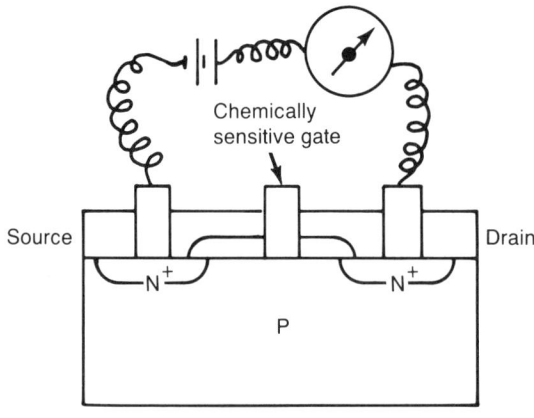

FIGURE 5.2
CHEMFET or ISFET Configuration
(Source: Weiss, 1985)

5.3
Fiber Optic Probes

Fiber optic probes for direct immersion in biochemical process fluids are technically feasible. These probes are low cost, sterilizable, rugged, corrosion resistant, and free from noise caused by electrical magnetic interference (emi) and radio frequency interference (rfi).

A pulsed fiber optic probe for cell density measurement in fermentors, shown in Figure 5.3, has been developed and tested. A light-emitting diode (LED) is driven by an oscillator circuit to provide monochromatic pulses. These pulses pass through a fiber optic line to a gap where it passes through the process fluid before entering another fiber optic line for transmission to a photodiode receiver. The photodiode output is connected to an ac-coupled amplifier where the dc component of the output that corresponds to the intensity of ambient light is eliminated. The attentuation of the intensity of the light pulses is exponentially related to the number of particles for a given probe gap geometry as shown in Figure 5.4 (Lee, 1981).

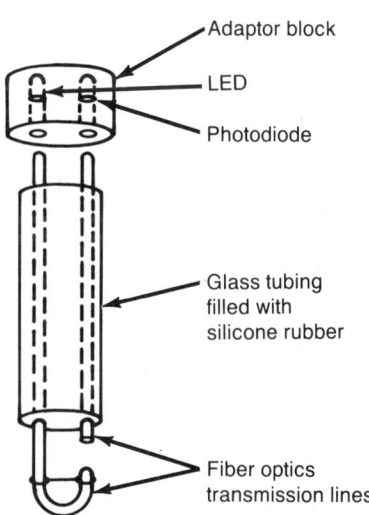

FIGURE 5.3
Fiber Optic Probe Construction
(Source: Lee, 1981)

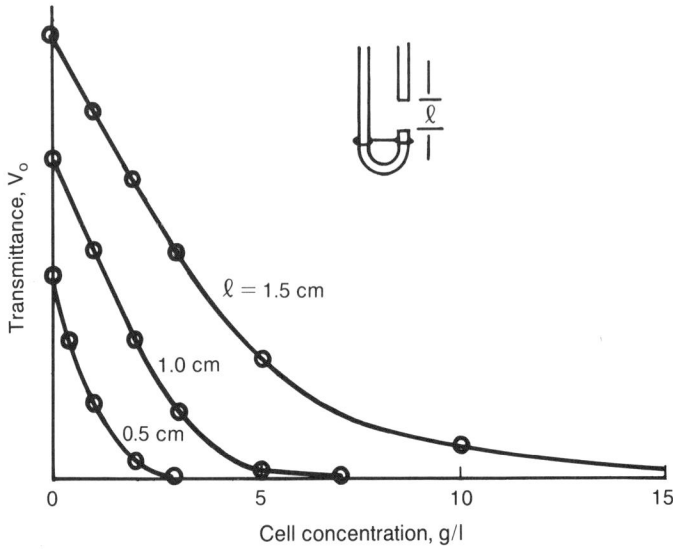

FIGURE 5.4
Fiber Optic Probe Output
(Source: Lee, 1981)

The probe naturally gets coated during the course of the fermentation. After 48 hours of an anerobic yeast fermentation, a calibration check of the probes showed that their output had decreased by five to ten percent. The decrease was less for probes with highly polished fiber optic ends. These probes are not suitable for mounting in aerated fermentors because the gas bubbles would cause too much noise. It may be possible to install them in the recirculation lines of loop fermentors and use electrical filters to screen out the effect of any remaining bubbles. However, other methods for monitoring light transmission or turbidity are available.

Guided Wave, Inc.® has developed a fiber optic probe for composition measurement using either ultraviolet, visible, or near-infared light. Light that is not absorbed by the process fluid is reflected from the end of the probe and transmitted by a fiber line to a direct-drive, servo-controlled holographic grating system for dispersing the light into its various wavelengths (for which patents are pending). This grating system is expected to have a continuous lifetime that is 20 times longer than that for conventional grating systems. The cost per point is reduced by use of an optical multiplexer and one analyzer. The display device is an IBM PC™ with color graphics (*Chemical Engineering*, 1985).

Fiber optic probes can be used for fluorometry, a method particularly suitable for the measurement of proteins, enzymes, vitamins, and pharmaceuticals. Fluorescense occurs when atoms or molecules are excited by the absorption of a beam of electromagnetic radiation and emit radiation upon return from the excited to the ground state. The emitted beam is most readily observable at a 90 degree angle to the excitation beam. The intensity of the observed fluorescense increases with the intensity of the incident radiation. As a result, fluorometry has a sensitivity that is two to four orders of magnitude better than that of corresponding spectrophotometric procedures. Tunable dye lasers can be used as the excitation source.

Cell culture fluorescence has been used as a measure of cell concentration. The measured fluorescence includes the effect of an environmental and metabolic fluorescence, as shown by Equation 5.3. The metabolic effect is rather large but has much faster dynamics than that of cell replication. Thus, it can be resolved as an additional piece of information from the composite signal. The metabolic component changes with operating conditions such as temperature, pH, dissolved oxygen, and substrate concentration. The environmental component has slow dynamics and is more difficult to resolve, but fortunately it is relatively fixed in comparison to the cell population component. Figure 5.5 shows the composite signal, the metabolic effects, the environmental effects, and the inferred cell concentration signal for a batch fermentation (Armiger, 1984).

$$F(t) = (Y_{F/X} \cdot (1 + M(t))) \cdot X(t) + E(t) \qquad (5.3)$$

where

$E(t)$ = environmental effect as function of time (NFU)
$F(t)$ = composite fluorescence as function of time (NFU)
$M(t)$ = metabolic effect as function of the time (dimensionless)
$X(t)$ = cell concentration (grams/liter)
$Y_{F/X}$ = fluorescence yield factor (NFU per grams/liter)

(NFU is normalized fluorescence units)

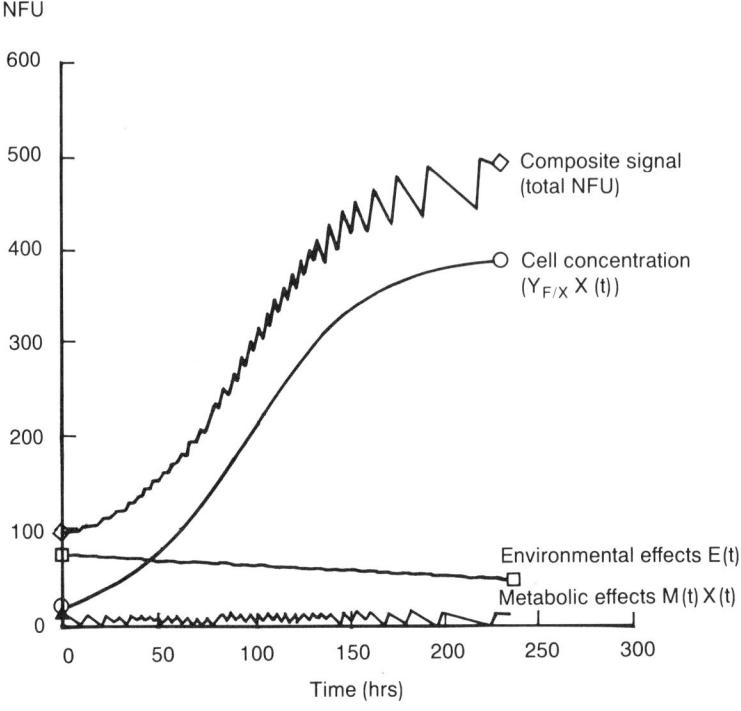

FIGURE 5.5
Cell Culture Fluorescence
(Source: Armiger, 1984)

5.4
Light Scattering

When light passes through a transparent medium, the light is scattered by molecules and particles or aggregates thereof. If the medium has particulate matter approaching the same dimension as the wavelength, the scattering is a measure of the turbidity of the solution. Turbidity has been used for many years as a measurement of cell concentration. The high cell concentrations achieved near the end of modern fermentation batches require that the broth sample be diluted. As an inference of cell concentration, it is not very accurate due to interferences. For example, protein accumulation within recombinant DNA E coli has been found to cause a large increase in the turbidity reading, as illustrated by Figures 5.6a and 5.6b (Fieschko, 1985).

COMPOSITION MEASUREMENTS

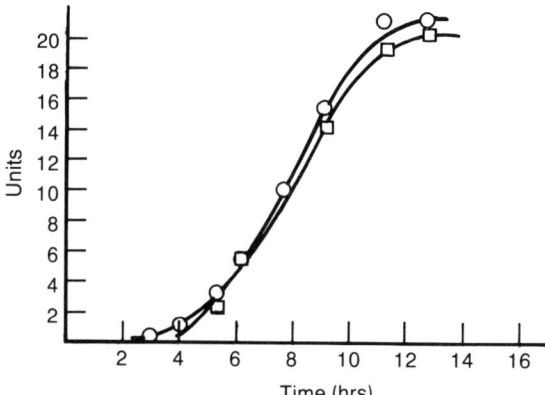

FIGURE 5.6a
Optical Density and Cell Dry Weight
(Source: Fieschko, 1985)

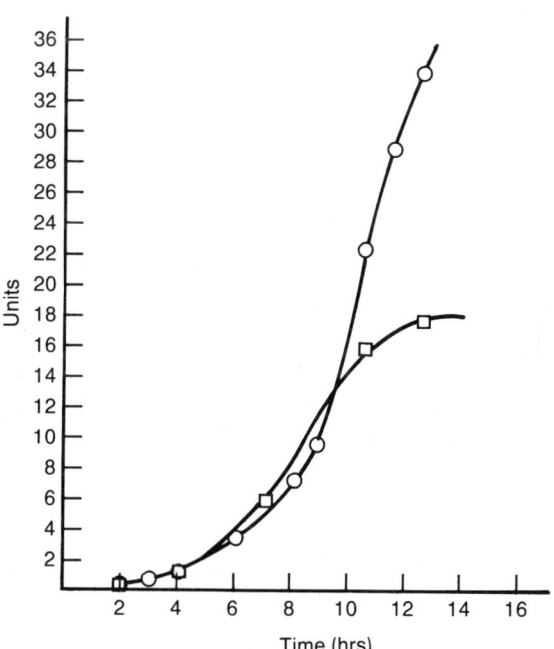

FIGURE 5.6b
Optical Density and Cell Dry Weight with Protein Accumulation
(Source: Fieschko, 1985)

If the scattering is done by molecules, some of the radiation undergoes a shift in wavelength as a result of inelastic collisions with the molecules. Since the total energy is conserved, the energy gained or lost by the photon must equal the energy change of the molecule. The shifted wavelengths provide a spectrum of the vibrational energy levels of the molecules. This method, called Raman scattering, provides as much spectral detail as infrared absorption methods without the major interference of water, a major constituent of biochemical streams. Raman and infrared methods have almost mirror-image spectrums so that much of the information cataloged for infrared spectrums can be transferred to Raman spectrums and vice versa. Since the amount of scattered light that has a wavelength shift is rather small, the intensity of the Raman bands is low. As a result, a tunable dye laser is used. When the laser wavelength lies under an intense electronic absorption band of a chomosphore (a functional group that contains valence electrons with relatively low excitation energies), enhancement of certain Raman bands occurs. Increases in Raman band absolute intensity of three to five orders of magnitude are achievable by this method, called resonance Raman spectroscopy. Furthermore, the bands can be selectively enhanced to cover the molecules of interest, which is a great advantage for extremely complex biological solutions (Carey, 1982).

Raman and resonance Raman spectroscopy are more generally used in the separation/purification area than in the reaction area because the methods are suitable for analyzing the product after it has been extracted from the broth. The sample must be free of particulates. Also, the sample must be kept moving through the laser fast enough to prevent the buildup of photo-products from biological transformation induced by the laser. This has been accomplished by use of a rotating or stirred sample sell. The minimum sample volume required is quite small (about 1 milliliter), which is an advantage for the analysis of high value-added purified products (Carey, 1982).

5.5
Nuclear Magnetic Resonance

Just as the absorption of ultraviolet or infrared radiation causes electronic transitions, the absorption of electromagnetic radiation of suitable frequencies causes transitions in the magnetic levels of nuclei created by a strong magnetic field. The molecular environment influences the absorption by the nucleus in the magnetic field so that the effect can be correlated with the molecular structure. The measurement is made by having a radio frequency source, magnetic field, and a radio frequency detector at right angles to each other. With no sample (no absorption), vector addition of the radio frequency components causes complete cancellation of the radio frequency in the direction of the detector. The addition of a sample causes incomplete cancellation (from the absorption of one vector of the frequency) and the appearance of a signal at the receiver. The strength of the coupling of the generator to the receiver depends upon the number of absorbing nuclei. A spectrum is generated by sweeping either the magnetic field or the radio frequency. The area under each peak in the spectrum is proportional to the quantity of the species present. The shift of the peak to a higher frequency with sample velocity can be used as an indication flow. The signal is independent of the sample phase (liquid, solid, or gas). However, the magnetic field within the sample must be homogeneous to one part in a billion. To achieve this stringent requirement, a phase lock system is used where a reference nucleus is irradiated, and any shift in the intensity of the frequency corresponding to its maximum resonance is used for feedback control of the magnetic field coils (Skoog, 1980).

Instruments are now commercially available, as shown in Figure 5.7, for the non-contacting and non-intrusive measurement of large scale process fluid flow and composition by transient nuclear magnetic resonance (NMR) analysis. The advantages of such construction in terms of meeting sterility and sanitary requirements of in-line biochemical measurements are obvious. However, its application to date has been in the area of proton NMR, whereas carbon-13 NMR is needed for the elucidation of biological structures. Carbon-13 NMR provides greater separation of peaks, narrower peaks, and fewer peaks (Skoog, 1985).

An analogous method that is also useful for biochemical applications is electron spin resonance (ESR). Many of the physical principles are the same, but microwave radiation is used instead of radio frequency radiation, and the absorption of radiation alters the spin of an unpaired electron instead of a nucleus. ESR is used for the determination of the concentration of paramagnetic ions in biological systems and the structural features, polarity, viscosity, and chemical reactivity of amino acids and functional groups in biological systems by the use of spin label reagents (Skoog, 1980).

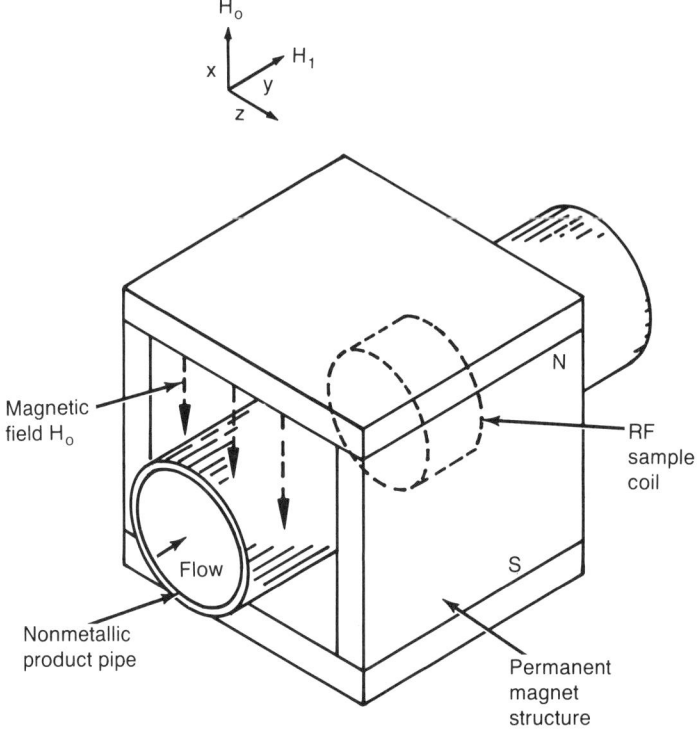

FIGURE 5.7
In-line Nuclear Magnetic Resonance Sensor
(Source: King, 1985)

5.6
Silicone Tubing Probes

Silicone tubing probes can be used to allow the selective diffusion of molecules in the broth for analysis in the gas pahse external to the fermentor. This is especially useful for dissolved oxygen measurements in mammalian cell cultures where the low velocities and long batch times would cause excessive coatings on conventional dissolved oxygen probes. Figure 5.8 shows the construction of silicone tubing probe for the measurement of methanol in the broth by the use of an external gas chromatograph. Volatile compounds in the fermentation permeate the

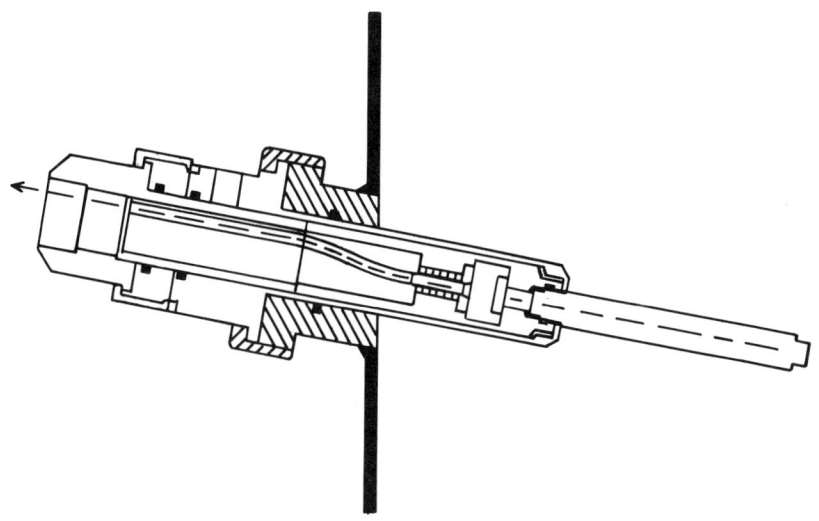

FIGURE 5.8
Silicone Tubing Probe for Fermentor
(Source: Kempe, 1983)

silicone membrane and are swept by a carrier gas to the analyzer. The speed of permeation depends upon the concentration of the compound and its permeation factor for the membrane. When used in combination with a mass spectrometer, it can provide accurate measurements for multiple fermentors (Kempe, 1983).

5.7
Mass Spectrometers

Mass spectrometers provide information on the concentration of different molecular weight species in a sample by the separation of an ionized gas sample in a magnetic field according to the species' mass-to-charge ratio. A small gas or volatilized liquid sample is ionized by bombardment by a finely focused electronic beam. Mostly positive ions are created by the stripping of an electron from the molecule. These positive ions are focused and accelerated into a strong magnetic field that causes the ions to trace a path whose curvature depends upon their mass-to-charge ratio. The ions strike Faraday collectors along the focal plane and pick up electrons to return to their neutral state. The resulting flow of electrons is converted by an electrometer into a usable voltage level. The position of the collector sets the molecular weight, and the voltage level determines the quantity of the species. An ion pump maintains a vacuum in the analyzer to minimize the collisions and scattering of ions. Figure 5.9 shows the internal configuration of an industrial unit (Perkin-Elmer, 1981).

A mass spectrometer has less drift, a faster response, greater accuracy, and fewer maintenance problems than the devices historically used for fermentor off-gas analysis. Within Monsanto, the paramagnetic analyzers used for oxygen measurement and the infrared analyzers used for carbon dioxide measurement in the off-gas served more often as examples of modern art than as functional devices after a few years of frustration. The performance of a mass spectrometer compared to these devices is like a Porsche compared to a go-cart. The advent of the microprocessor

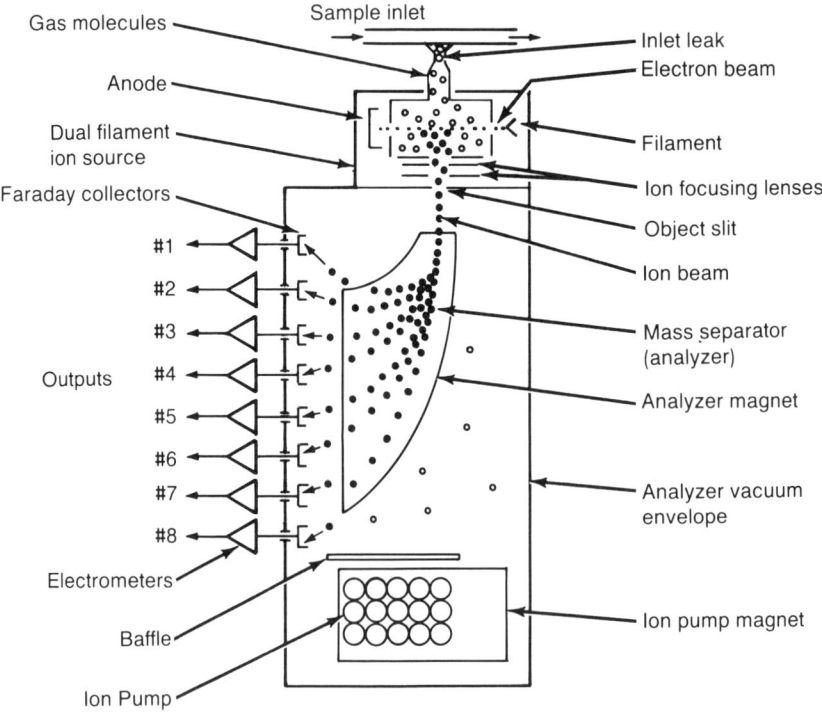

FIGURE 5.9
Internal Configuration of a Mass Spectrometer
(Source: Perkin-Elmer, 1981)

helped drop the price and reduce the complexity of adjustments of the mass spectrometer to the point where it is the preferred method of off-gas analysis. Because it can make an analysis in a few seconds, one mass spectrometer can service ten or more fermentors. Besides oxygen measurement for calculation of the oxygen uptake rate and carbon dioxide measurement for calculation of the carbon dioxide production rate, many other specie measurements are also available. For example, an inert component in the air feed can be measured in the off-gas to provide a check of the air flow meter. If nitrogen is not consumed in the reaction in the fermentor, it is used otherwise as an inert gas (such as argon) and is injected in the air feed and used as the trace component. In general, the mass spectrometer can provide an accurate material balance for a batch fermentor by off-gas analysis (what isn't going out the vent must be left in the batch). Cell metabolism, cell concentration, and broth composition can be inferred. It can be used with a computer system for operator advisory control and in conjuction with growth models for feedback supervisory control (see Chapter 7 for more details).

CHAPTER 6

Control Systems

Technology has taken another giant step in the wrong direction. Most of the literature to date on biochemical control systems is from the universities whose forte is the periodic invention of a new algorithm for feedback control. These algorithms have impressive names such as optimizers, predictors, observers, and compensators and have formidable mathematical bases. If laid end to end, all the equations for new algorithms published in the last ten years would stretch to the moon and back. While the mental gymnastics are beneficial in terms of improving one's intellectual flexibility and dexterity, they serve as diversions from the real issues. The result is a very confused state (McMillan, 1986).

Despite claims to the contrary, advanced control algorithms cannot do better than a well-tuned conventional proportional-plus-integral-plus-derivative (PID) controller for unmeasured load disturbances. Studies that show better performance from a new algorithm typically omit derivative action, or have less than the best tuning settings for the conventional controller, or set point instead of load disturbances. While these algorithms cannot outperform a conventional controller on a single-loop basis, they can be designed to include decouplers for interacting loops and feedforward control for measured disturbances. Also, given the loop dynamics, the methods for calculating the controller tuning settings are more accurate. This all sounds important until you realize that for biochemcial control loops, interaction is a minimal problem, measured disturbances are rare, and only approximations of the process and instrument dynamics are known. The key to improving a loop's performance unfortunately is not as academically clean as incorporating a new advanced control algorithm but involves obtaining the best sensor and final element and reducing the loop dead time (McMillan, 1986).

The major contributors to the total error between the measurement and the set point are the measurement error, measurement noise band, final element precision, and the control error from load disturbances. As to the size of control error as a contributor, it is relatively small in the reaction/conversion area and relatively large in the separation/purification area.

Since the biochemical process control field is in its infancy, much work still remains in merely obtaining better measurements. As seen from the previous chapters, it is not easy to come up with sensors that will be specific and accurate enough and survive in an environment that alternates between slime and steam sterilization. Where sensors have not been developed, inferred measurements are used. For example, calculations like the oxygen uptake rate (OUR) and carbon dioxide production rate (CPR) are made from off-gas measurements to provide inferred measurements of broth cell concentration. Sometimes it requires a little bit of imagination to see the correlation between the inferred measurements and the actual process variables, unless the biochemistry is understood enough to get an approximate model.

Even if one ignores measurement noise from electrical magnetic interference (emi) or radio frequency interference (rfi), there is still plenty of noise in biochemical loops. The culprit is the equipment or piping. Flow measurement noise is a way of life for most flow loops and liquid pressure loops, due to pressure transients from unsteady flow around obstructions, shifting velocity profiles, and positive displacement pumps. Level measurement noise in agitated or aerated vessels is common. Composition measurement noise occurs in the reaction/conversion from air bubbles and in the separation/purification area from in-line equipment that provides little back-mixing. Temperature measurement loops are the only loops one can count on as being noise-free because of the slowness of the process and sensor response. Measurement uncertainty from noise not only contributes to the total error but can cause excessive wear of the final element. To eliminate this concern, the controller proportional band is increased (gain is decreased) or a measurement filter time constant is added. The goal is to attenuate the oscillation amplitude of the controller output to the point where it lies within the dead band of the final element.

Equation 6.1 shows how to estimate the proportional band required, and Equation 6.2 shows how to estimate the filter time constant required. As usual, you don't get something for nothing. The penalty you pay is a larger control error. The larger proportional band causes larger control errors because of less control action, and the filter time constant causes larger control errors because of delayed sensing of true load disturbances. The larger proportional band tends to increase the peak error more than the integrated error from a load disturbance, where the reverse is true for the filter time constant since it increases the period of the loop (McMillan, 1983).

$$PB = 100\% \cdot \frac{E_p/8}{E_v} \cdot \sqrt{\frac{0.8 \cdot T/N}{TC_m + 0.8 \cdot T/N}} \quad (6.1)$$

$$TC_m = \frac{E_p \cdot 0.5 \cdot T/N}{E_v \cdot 2\pi} \quad (6.2)$$

where

E_p = peak amplitude of process error oscillations (%)
E_v = dead band of final element (%)
PB = proportional band of controller (%)
N = number of zero crossings of process error (dimensionless)
T = time interval of N zero crossings (minutes)
TC_m = measurement filter time constant (minutes)

All mechanical components have a dead band or backlash effect where a change in controller output signal results in no movement of the component. It occurs whenever the motion of the component has been at rest or changes direction. In globe valves, it is mostly caused by the stiction of the stem packing, and in rotary valves it is also caused by the stiction of the ball or disc seal and the slop in the linkages. Since sanitary control valves cannot have stem packing, these types of valves are not generally used in biochemical control loops. Until recently, the Grinnel-Saunders® diaphragm valve was the one type of control valve that met the construction requirements for sanitary operation. The only dead band in these valves is that associated with the filling and exhausting of actuator air and the stiffness of the diaphragm, which changes with the number of sterilizations with live steam. However, the inherent valve flow characteristic is somewhat irregular and the ability to throttle at low lifts is rather limited. Also, the minimum size has too much capacity for reagent and nutrient feeds except for large fermentors. H. D. Baumann Assoc., Ltd. has developed a packless control valve, shown in Figure 6.1, that has a unique mechanical amplifying mechanism to

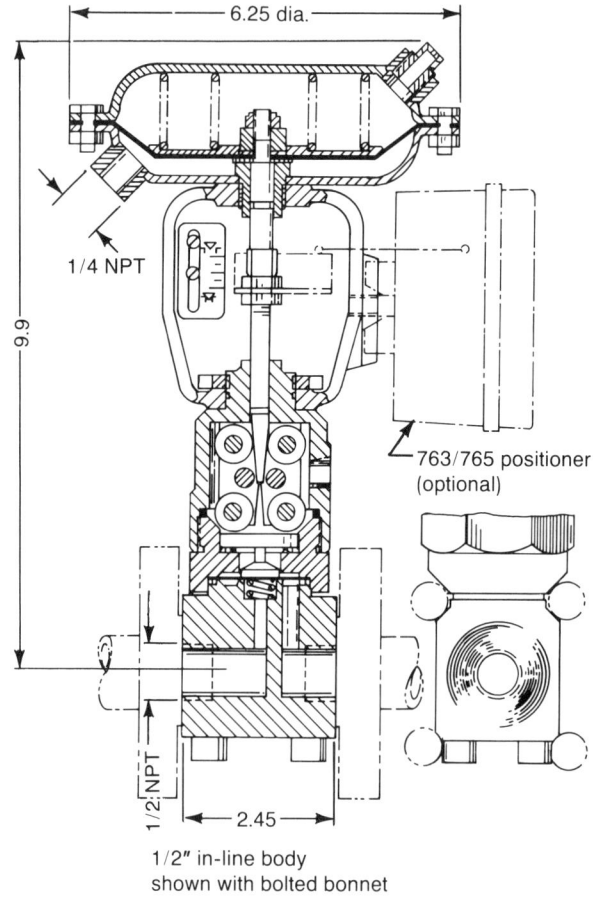

FIGURE 6.1
Packless Sanitary Control Valve
(Source: H. D. Baumann Assoc., Ltd.)

CONTROL SYSTEMS

precisely position a nickel alloy or Hastelloy C™ diaphragm against an orifice-to-throttle flow. The valve has tremendous rangeability (1000:1), low flow coefficients ($C_v = 0.03$ to 0.7), and negligible dead band. The valve comes with either a bolted or quick disconnect bonnet that when removed provides access to the valve seat and closure diaphragm for cleaning and inspection while the valve stays in the line (the actuator is attached to the bonnet).

One point often overlooked is the dead band in the manipulated variable created by the signal gap between split-ranged final elements. The gap is purposely included in the loop to eliminate the simultaneous opening of both final elements. For pH control,

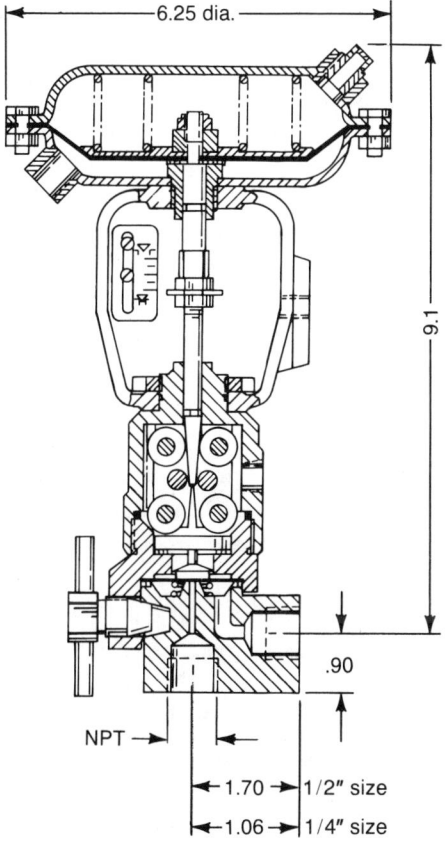

1/4" & 1/2" angle body
shown with quick disconnect bonnet

FIGURE 6.1 (continued)
Packless Sanitary Control Valve
(Source: H. D. Baumann Assoc., Ltd.)

the gap prevents the simultaneous use of both acid and base reagents to reduce reagent consumption. For tempered water temperature control, the gap prevents the simultaneous use of both chilled water and steam to produce energy consumption. For dissolved oxygen control, a gap between the split range of air feed and agitator speed serves no such purpose and should not be used. While the gap reduces reagent and energy for competing final elements, the loop performance deteriorates. The dead band creates an additional dead time that is proportional to the gap width and inversely proportional to the speed of the controller output. The larger dead band causes one to detune the controller (less gain and less reset action), which causes the controller response to be more sluggish, which increases the dead time. The whole thing can snowball. The gap can also cause sustained oscillations when there is no reagent or utility demand, because any slight offset of the measurement from the set point will be used by reset action to drive the output back and forth across the gap. Thus, the gap should be as narrow as possible. Towards this end, the gap should be computed in a microprocessor-based controller rather than calibrated into a positioner for the valves because of the greater gap accuracy at the onset, nonexistent drift, easier gap verification, and easier gap modification. There is a limit to how much the gap can be narrowed. All control valves require more signal to open than to close from friction of the seating surfaces. The gap cannot be smaller than this dead band between the open and closed signals.

Control error occurs because there is dead time in control loops. In order for an equal but opposite compensating action to cancel the effect of an unmeasured load disturbance, the disturbance must propagate through the process, be sensed and transmitted to the controller, be recognized and acted on by the controller, initiate a change in the final element, and the manipulated variable propagated through the process to the point of disturbance entry. In other words, the signal must make one complete pass around the control loop, as shown in Figure 6.2, before an excursion in the controlled variable from an unmeasured load disturbance can be arrested. Thus, the peak error occurs after one loop dead time. The excursion rate of the process variable is set by the process and disturbance time constants. To minimize the peak error, the process time constant should be maximized for self-regulating processes, the dead time minimized, and the disturbance time constant maximized for all processes. A strange but true twist to these rules that will be explained

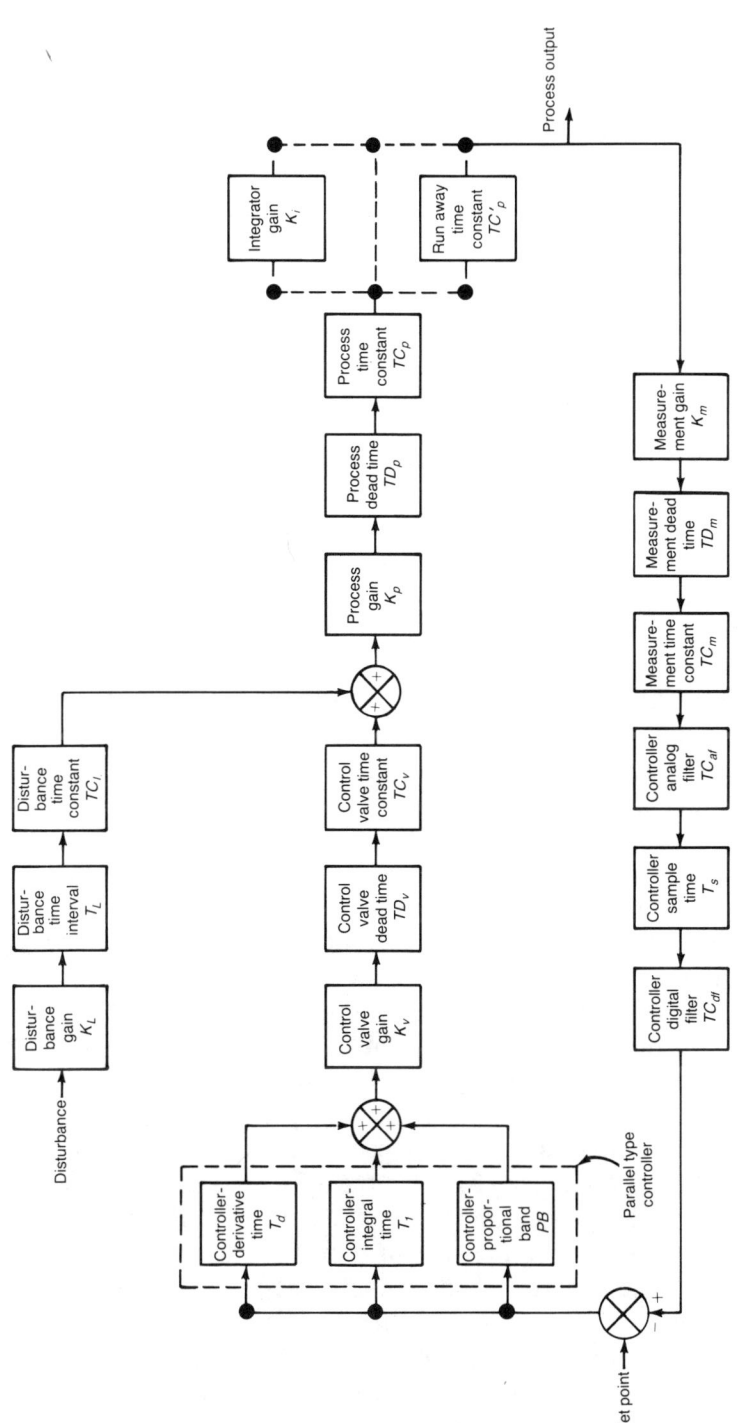

FIGURE 6.2
Block Diagram of a Generic Control Loop

subsequently is that the process time constant should be minimized for nonself-regulating (integrating and runaway) processes. These rules apply only if the controller has the best possible tuning settings. If the controller is poorly tuned, it doesn't matter how nice the loop dynamics are, the performance will be lousy. Taken to the extreme, as the reaction of the controller becomes nil, the closed-loop response approaches the open-loop response (McMillan, 1983).

6.1
Error Prediction

If you are not into equations for detailed error prediction, you might take this opportunity to go to the refrigerator, grab yourself a brew (biochemically produced, of course), relax, and skim through this section. It is like taking an interstate bypass around a busy and complex city with only the silhouette of the skyscrapers seen on the horizon. Just remember to get off at the next section.

The peak error is the maximum deviation of the controlled variable from the set point. Control theory textbooks don't talk about it much, even though it is typically the most important control error, because it cannot be predicted rigorously. However, the fraction of the open-loop error (controller on manual), seen as the closed-loop error (controller on automatic), can be approximated by Equation 6.3. For small controller proportional bands, the proportional band in the denominator can be dropped so that the peak error is proportional to the controller proportional band (inversely proportional to the controller gain). Not evident yet from the equation is that the gain terms have no effect on the portion of the open error seen as the peak error because the controller proportional band is proportional to the same gain product if properly tuned. However, the open-loop error itself is proportional to the process gain. Thus, an increase in the valve or measurement gain increases the proportional band required, but the absolute peak error remains about the same. An increase in the process gain increases both the proportional band required and the absolute peak error. It is important to realize that an increase in measurement or valve gain will probably increase the measurement and final element errors whenever the higher gains are due to a larger measurement span and larger valve, because accuracy and valve hysteresis are typically a percent of full scale reading or stroke (the full scale stroke translates to a larger flow for a larger valve).

$$E_x = \frac{K \cdot PB}{100\% \cdot K_v \cdot K_p \cdot K_m + PB} \cdot E_o \qquad (6.3)$$

$$E_i = \frac{PB}{100\% \cdot K_v \cdot K_p \cdot K_m} \cdot T_i \cdot E_o \qquad (6.4)$$

where

E_x = peak error (%)
E_o = open-loop error (%)
K = constant depending upon controller type and degree of damping (K = 1.1 for most PID controllers and quarter amplitude damping)
K_v = valve (final element) gain (flow units/% signal)
K_p = process gain (process units/flow units)
K_m = measurement gain (% signal/process units)
PB = controller proportional band (%)
T_i = reset time of controller (minutes/repeat)

The integrated error is the total of the deviations of the measurement from the set point. On a plot of the closed response of a loop to a load disturbance, it is the net area between the measurement oscillation and the set point. It is also known as the accumulated error. For a composition control loop, the integrated or accumulated error is an indication of the net surplus or deficit of a particular component. The integrated error is more appropriate than the integrated absolute error because the short-term plus and minus excursions in composition cancel out from the filtering effect of process equipment capacity, unless there is a nonreversable reaction. The integrated or accumulated error is proportional to the product of the controller proportional band (the inverse of controller gain) and integral time (the inverse of controller reset), as shown in Equation 6.4. For a smaller proportional band, the amplitude of the oscillation is less, and for a smaller integral time, the period of the oscillation is less (McMillan, 1983).

The equations for the peak and integrated errors lead to several conclusions important enough to emphasize. First, if the disturbances are nearly zero in magnitude (open-loop error approaches zero), even the most difficult loop will perform excellently. Thus, before one can decide whether a difficult loop justifies the additional expense of special equipment or instruments, knowledge of the size of the disturbance is necessary. Second, if the controller was tuned with too large a proportional band (too little gain) or too large an integral time (too little reset action), an easy loop will perform as poorly as a more difficult loop that requires the higher mode settings. Any special efforts or expense during design to improve loop performance will be wasted if overly conservative controller tuning is used in the field. Third, if the resolution or rangeability of the mode settings of the controller prevents the use of both the best proportional band and reset settings, an easy loop will again perform as poorly as the more difficult loop. Any capital expended for hardware or design to improve loop performance is wasted where proportional band or integral time settings are required that are below the lower limit of the available controller. Fourth, if the process gain is large, the open-loop error and, hence, the closed-loop error will be large. It is, therefore, important that the process engineer and the control engineer review the effect of process equipment and operating conditions on the process gain at an early stage of the project.

The ISA Monograph, *Tuning and Control Loop Performance*, has more equations than an engineer should be allowed to have to predict the controller tuning settings. To repeat the equations in all their gory detail would run the risk of failure to see the forest because of all the trees. Instead, the method will be discussed in one marathon paragraph and the conclusions will be summarized in Table 6.1. For those readers who cannot walk away from an academic challenge, Appendix A shows how these equations are used to predict the control errors for composition control of a biochemical reactor in the exponential growth phase.

The equations require approximations of the loop dead time, the loop time constant, and the open-loop gain. The loop dead time is estimated by summing all the pure time delays with all the equivalent time delays from the small time constants in the loop. The loop time constant is estimated as the summation of the fractions of the loop time constants not converted to dead time. A simpler and more conservative approach is to use the largest time constant in the loop as the loop time constant. The open-loop gain is the final change in controller measurement input signal divided by the change in the controller output signal. The input and output signals should be converted to percent so that the quotient is dimensionless. For an integrating process, the open-loop gain

TABLE 6.1
Conclusions from Equations to Predict Tuning Settings

Process type	Loop condition	Conclusion
Self-regulating	Ratio of loop dead time to time constant is small	Peak error is approximately ratio times open-loop error
Self-regulating	Ratio of loop dead time to time constant is small	Peak error is proportional to dead time
Self-regulating	Ratio of loop dead time to time constant is small	Integ. error is proportional to the dead time squared
Self-regulating	Ratio of loop dead time to time constant is small	Ultimate PB is approximately 2/3 of 100% times ratio
Self-regulating	Ratio of loop dead time to time constant is small	Ultimate period approaches four times the dead time
Self-regulating	Ratio of loop dead time to time constant is large	Ultimate PB is approximately 100% times open-loop gain
Self-regulating	Ratio of loop dead time to time constant is large	Ultimate period approaches two times the dead time
Integrating	Ratio of loop dead time to time constant is large	Ultimate period approaches four times the dead time
Integrating	PB used is 100 times larger than best PB calculated	Process variable develops nearly sustained oscillation
Runaway	Ratio of loop dead time to time constant is large	Ultimate period approaches four times the dead time
Runaway	Loop time constant or dead time approaches process positive feedback time constant	Ultimate period approaches infinity and window of allowable PB closes
Runaway	PB used is larger than 80% times open-loop gain	Process variable diverges from set point

includes the integrator gain, which is a ramp rate (the open-loop error is a ramp). For a runaway process, a positive feedback time constant for the process must be estimated. These parameters are used to approximate the ultimate period and ultimate proportional band for closed-loop operation. The Ziegler-Nichols equations for the closed-loop tuning method are then used with a calculated rather than a measured ultimate period and ultimate proportional band. The Ziegler-Nichols closed-loop method gives smaller control errors than the Ziegler-Nichols open-loop method and other documented tuning methods for PID controllers. (McMillan, 1983).

Nearly all of the literature to date on process control has focused on self-regulating processes even though the type of process affects the tuning settings, as shown in Table 6.1, and the performance of the loop. Table 6.2 classifies some of the more common biochemical loops as to their process type. The pseudo-integrating loops are not true integrators, but they behave as such because their process time constant is so large that the open-loop response looks like a ramp within the control band. For integrating and runaway processes, the largest negative feedback constant slows down the correction for a diverging process variable. Whereas this time constant was beneficial for self-regulating processes, it is now detrimental and should be minimized along with the loop dead time.

TABLE 6.2
Process Types for Biochemical Loops

Loop type	Operating condition	Process type
Flow	—	Self-regulating
Liquid pressure	—	Self-regulating
Gas pressure	Small volume/flow ratio	Self-regulating
Gas pressure	Large volume/flow ratio	Pseudo-integrating
Gas pressure	Dead-ended volume	Integrating
Level	Pumped-out discharge	Pseudo-integrating
Level	No discharge (batch)	Integrating
Level	Gravity flow discharge	Self-regulating
Dissolved oxygen	Batch with off-gas flow	Self-regulating
Dissolved oxygen	Batch without vent	Integrating
Dissolved oxygen	Continuous	Self-regulating
Temperature	Batch or continuous	Self-regulating
pH	Batch	Integrating
pH	Continuous	Self-regulating
Substrate or product concentration	Batch	Integrating
Substrate or product concentration	Continuous	Self-regulating
Cell concentration	Exponential growth phase	Runaway
Cell concentration	Batch and stationary phase	Integrating
Cell concentration	Continuous and stationary phase	Self-regulating

6.2
Effect of Process Equipment Type

The effect of process equipment naturally depends upon the type of loop. For flow and liquid pressure loops, the process equipment has little to no effect since the total error in these loops is largely determined by the speed and accuracy of the instruments and the degree of measurement noise. For level loops, the vessel diameter greatly affects the control error because the integrator gain and, hence, the open-loop error are inversely proportional to the square of the diameter.

For a given height, the diameter of a vessel has the same effect on composition loops as for level loops since the back-mixed volume sets the loop time constant for continuous operation and the integrator gain for batch operation. However, the turnover time for back-mixed regions and the transportation time for plug flow regions are major contributors to the loop dead time. Both of these increase with the volume. Thus, while a larger vessel may increase the loop time constant or decrease the integrator gain, it will also increase the loop dead time. The integrated error increases rapidly with vessel size since it is proportional to the dead time squared. The peak error would theoretically stay the same if the agitator pumping rate for the back-mixed region and the throughput rate for the plug flow region were scaled up with the volume. Practically speaking, this rarely occurs because the pumping power costs increase with volume to the third power. Even if you could afford the power costs, the fluid flow patterns deteriorate (stagnation areas develop and path lengths increase). Figure 6.3 shows how a composition control loop would perform for the same disturbance if installed in a well-mixed vessel, a poorly mixed vessel, and an in-line system. Note that the peak error of the in-line system is the largest (it is essentially the open-loop error), and the peak error for the well mixed tank is the smallest. The in-line system has the smallest integrated error. This feature is attractive if the loop can be stabilized (the open-loop gain is not too high) and a volume exists downstream to attenuate the amplitude of the oscillations. The poorly mixed tank performance is unacceptable because the integrated error is the largest, and the period of oscillation is too large to facilitate filtering of the relatively large peak error.

For temperature loops that manipulate a heat transfer stream instead of a feed stream for inferred composition control, the

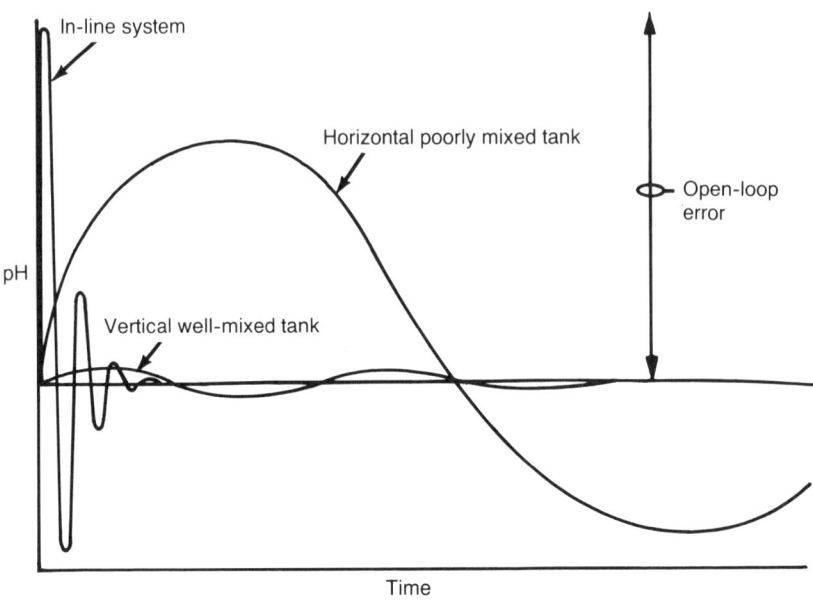

FIGURE 6.3
The Effect of Equipment Type on Composition Control

volume has the same effect, but mixing is less a factor because of rapid heat dispersion and equalization. However, other equipment design features are now important. The heat transfer area and coefficients determine a thermal lag that is similar in effect to a very large valve time constant. The volume of the coils or jacket create additional dead time that is similar in effect to a very large valve prestroke dead time. Both of these terms slow down the ability of the manipulated variable to counteract a load disturbance.

6.3
Critical Loop Characteristics

Since a picture is worth a thousand words and much easier on the eyes, plots of the important controlled and manipulated variables for a fed batch fermentation are shown in Figures 6.4 through 6.12. The plots are the output from an Advanced Continuous Simulation Language (ACSL) program written to study the effect of loop dynamics on batch and continuous fermentor control. The program includes the effect of bubble dynamics on oxygen transfer, the effect of temperature on the maximum growth rate and cell maintenance coefficients, and the effect of pH from an ion charge balance on the maximum growth rate and the dissolved carbon dioxide concentration. The pH reagent requirement was increased with cell population. In retrospect, it should have also been made a function of the dissolved oxygen concentration. The program listing is documented in Appendix B. The program output plots show how the increased turnover time of a loop fermentor, compared to that of a stirred fermentor, increases the control error (particularly the integrated error) and how this can be offset by faster measurements.

Perhaps the toughest and most critical loop for aerobic fermentations is the dissolved oxygen loop. The process dead time is larger than for other composition control loops because it includes the rise time of the bubbles as well as the turnover time of the mixing. While this causes a deterioration in loop performance, its effect is overshadowed by problems with measurement noise and the manipulated variable.

The temporary attachment of gas bubbles to the diaphragm alters the area for transport across the membrane enough to cause the reading to change. The effect is particularly noticeable for galvanic probes because their reading for oxygen in the gas is about half that for oxygen in the liquid. The frequency of the noise depends upon the type and amount of anti-foam. A larger number of smaller sized bubbles, which want to adhere more closely the membrane, causes the frequency to decrease. At the end of a batch run, the frequency is so low it resembles dc drift. The addition of anti-foam agents under on-off control also causes rapid spikes in the dissolved oxygen that can be considered to be noise since they are too fast to be controlled. Finally, the process gain is also larger than for other fermentor composition loops so that short-term concentration gradients have a greater effect on the reading.

The real culprit is the manipulated variable. The effectiveness of the manipulated variable and, hence, its gain changes with the mass transfer coefficient between the gas and liquid phases, which is a complex function of agitation and the physical properties of the broth. Dissolved oxygen control from the beginning to the end of the batch requires gas flow rangeability beyond the capability of a conventional air flow control valve, not to mention a diaphragm valve. Single-step batch fermentations typically do not

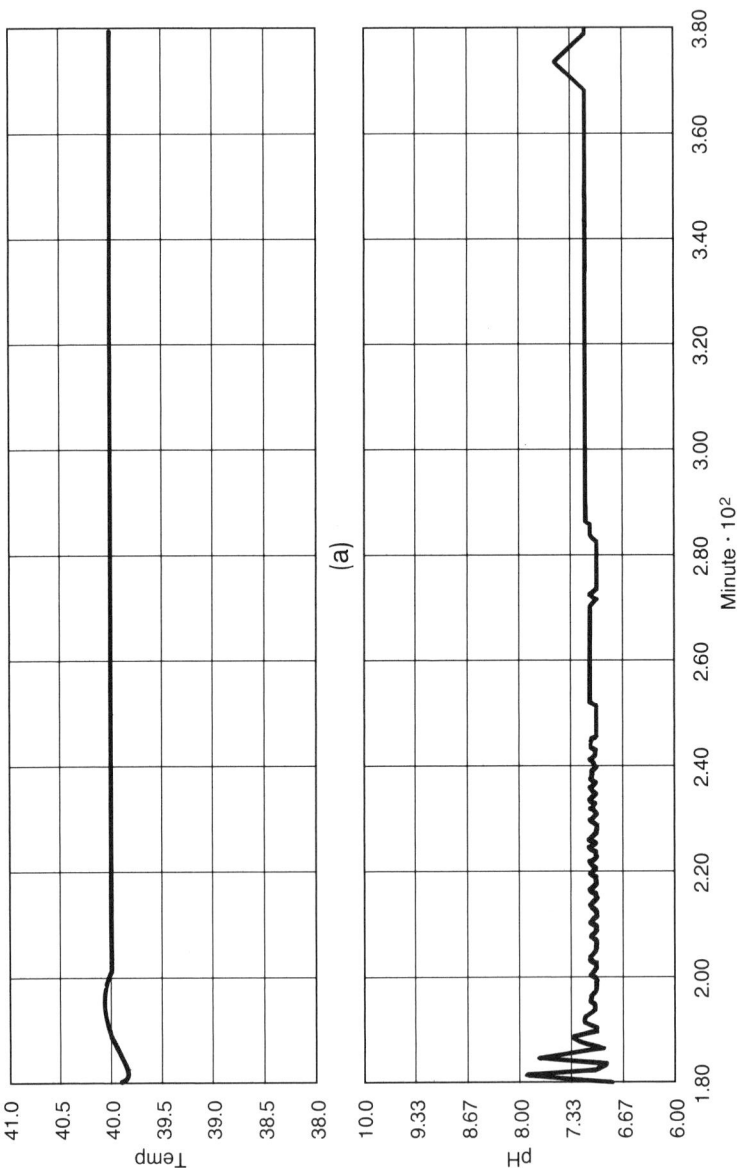

FIGURE 6.4
Fed Batch 1500 Liters Stirred Fermentor/Fermentor pH (pH)/Temperature (Temp)

CONTROL SYSTEMS (111)

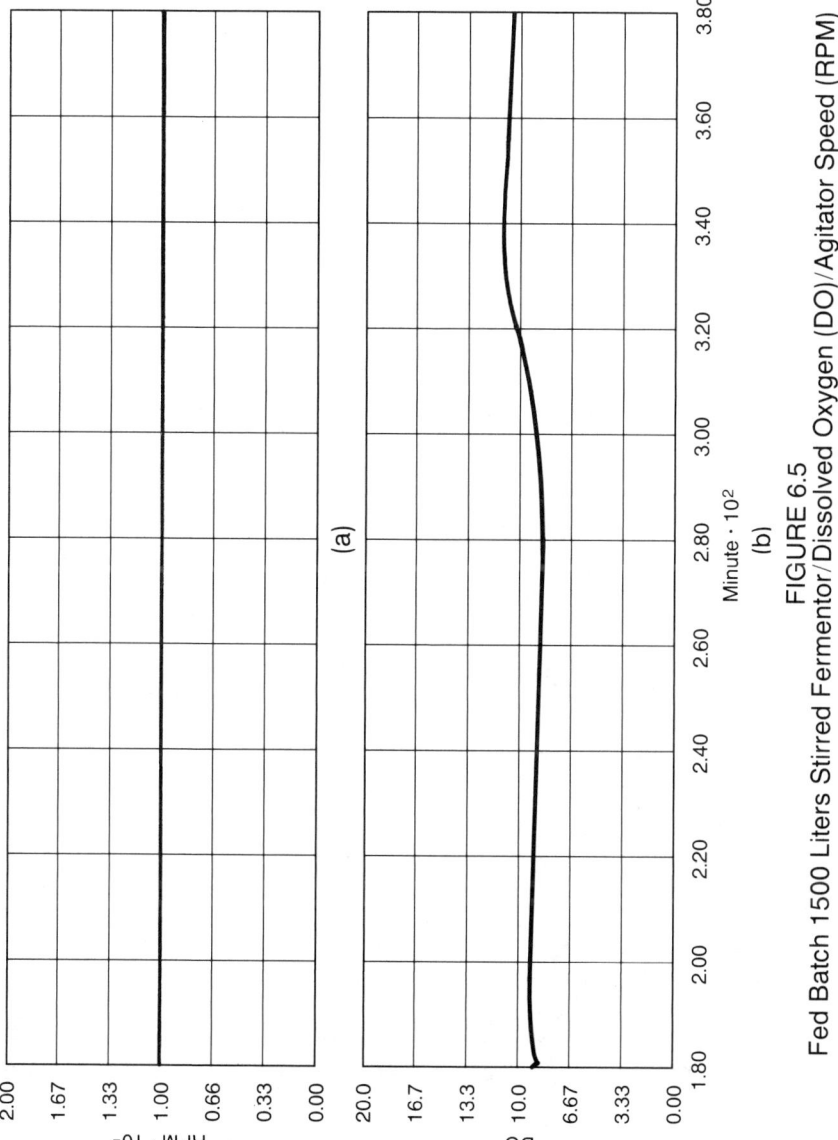

FIGURE 6.5
Fed Batch 1500 Liters Stirred Fermentor/Dissolved Oxygen (DO)/Agitator Speed (RPM)

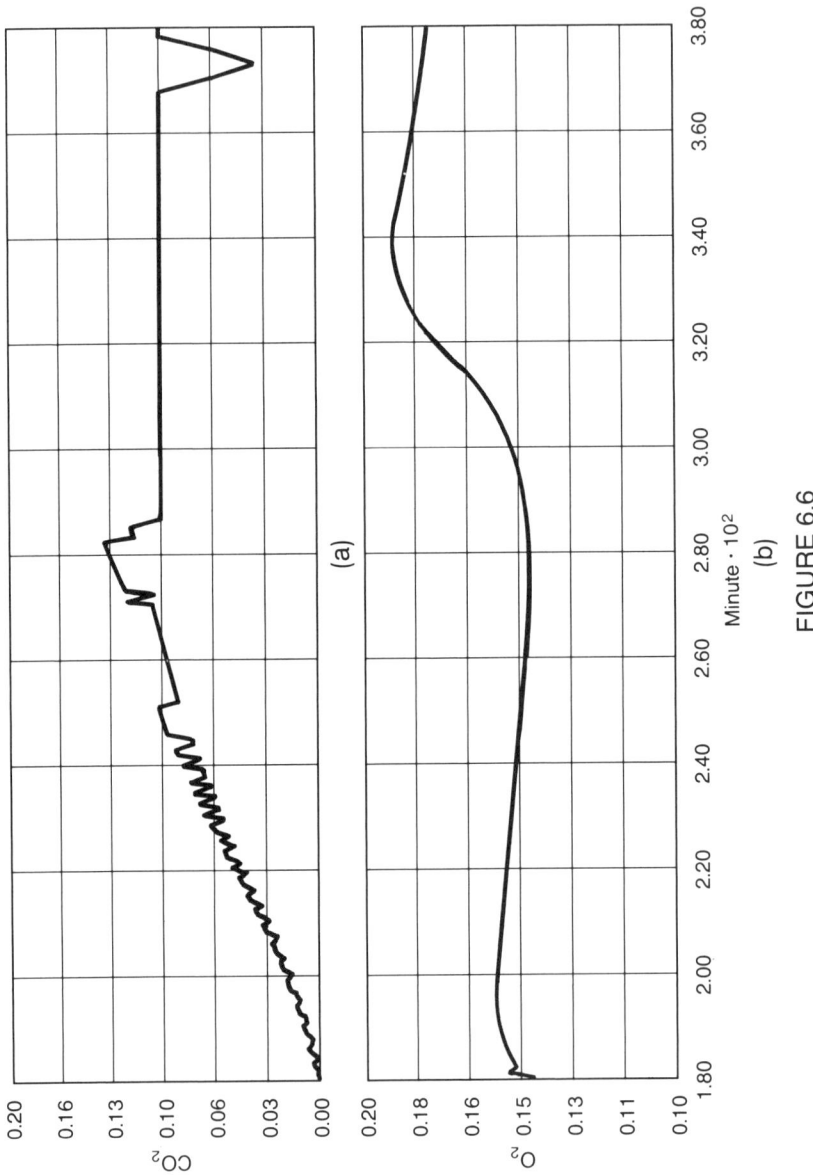

FIGURE 6.6
Fed Batch 1500 Liters Stirred Fermentor/Oxygen in Off-gas (O_2)/Carbon Dioxide in Off-gas (CO_2)

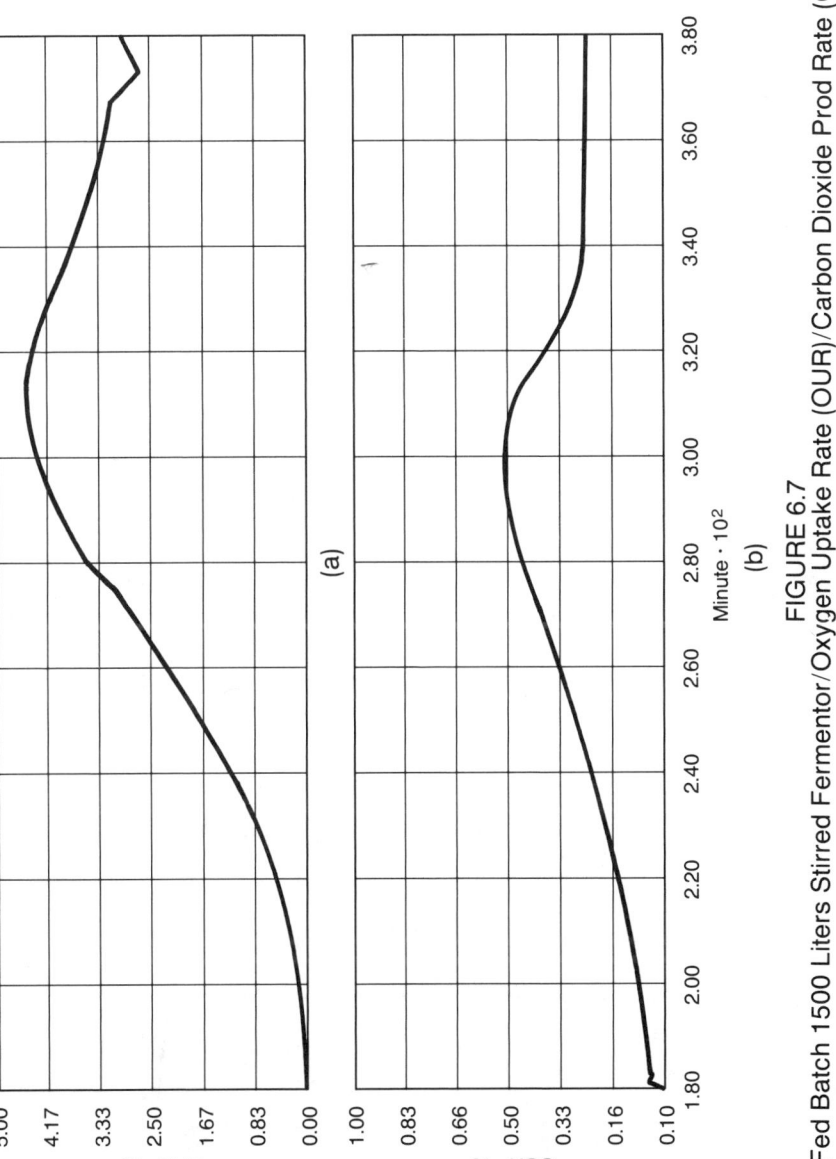

FIGURE 6.7
Fed Batch 1500 Liters Stirred Fermentor/Oxygen Uptake Rate (OUR)/Carbon Dioxide Prod Rate (CPR)

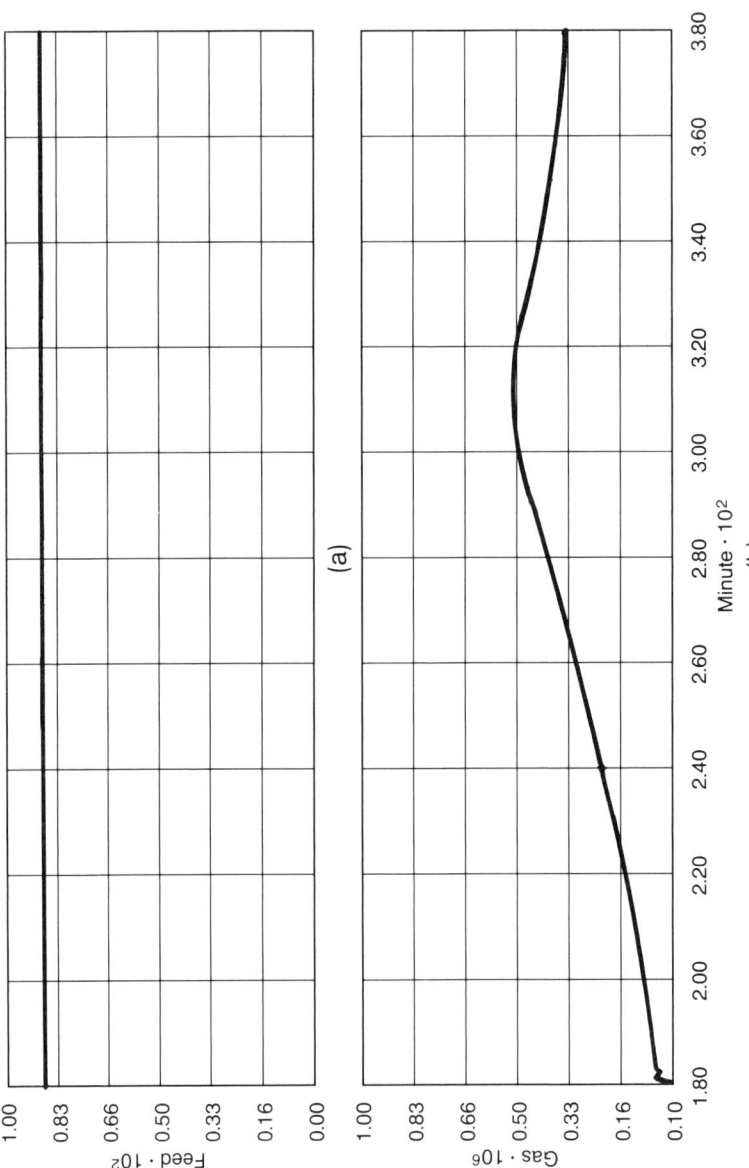

FIGURE 6.8
Fed Batch 1500 Liters Stirred Fermentor/Gas Flow (Gas)/Substrate Feed (Feed)

CONTROL SYSTEMS

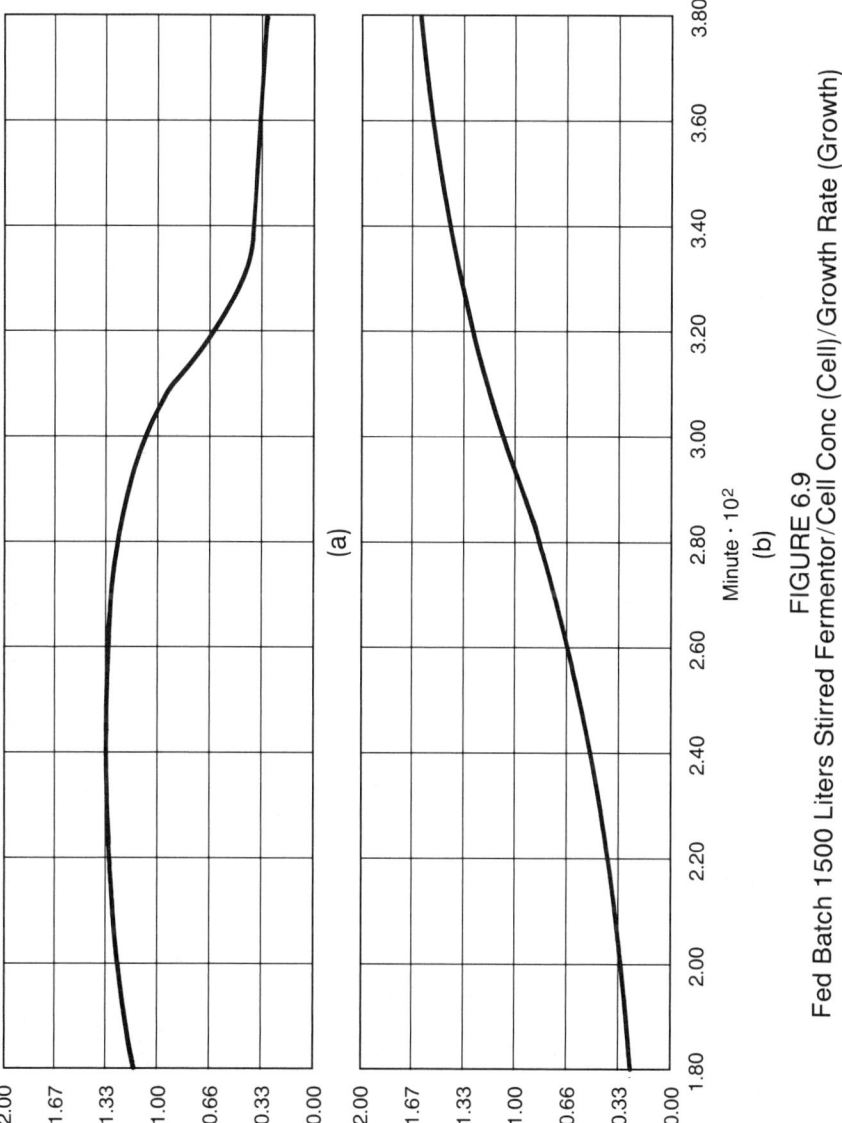

FIGURE 6.9
Fed Batch 1500 Liters Stirred Fermentor/Cell Conc (Cell)/Growth Rate (Growth)

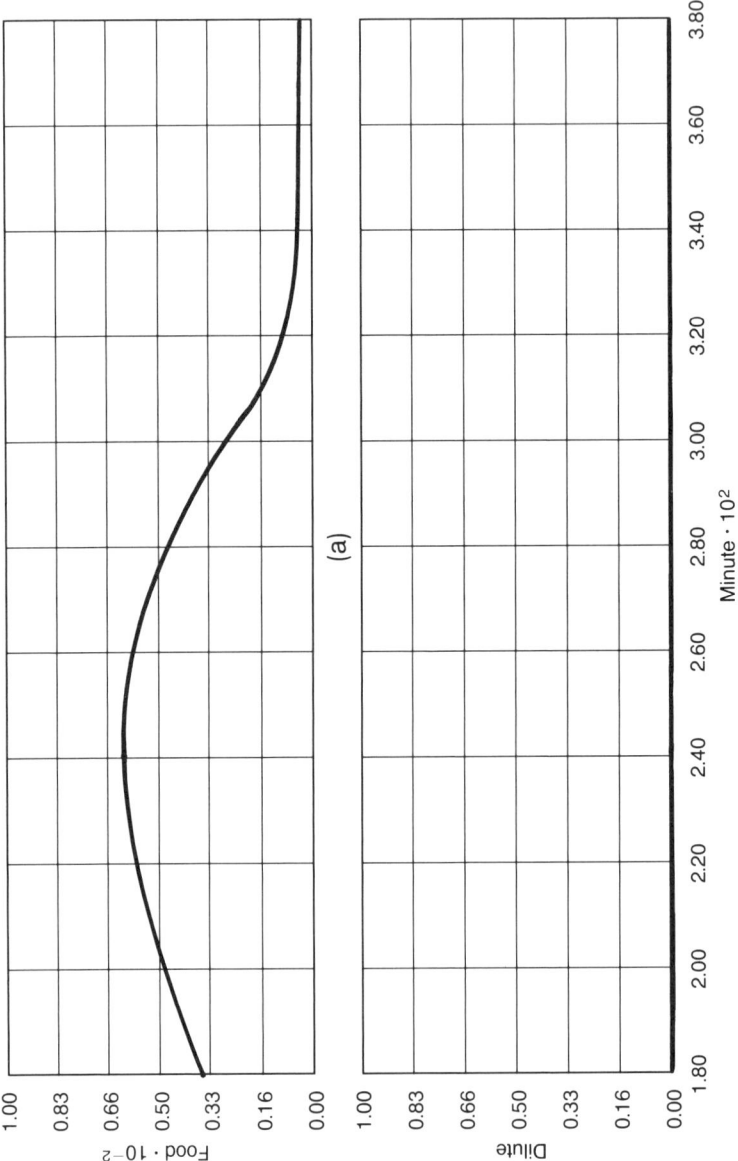

FIGURE 6.10
Fed Batch 1500 Liters Stirred Fermentor/Dilution Flow (Dilute)/Feed Conc (Food)

CONTROL SYSTEMS

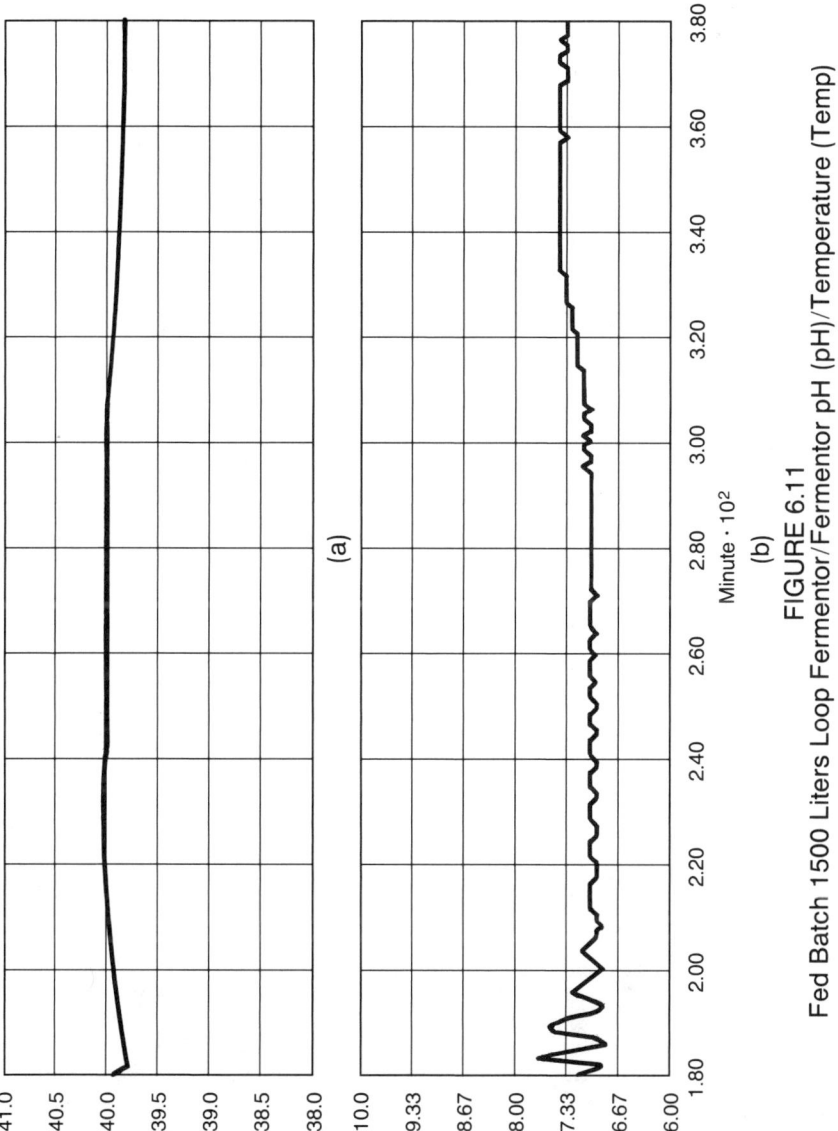

FIGURE 6.11
Fed Batch 1500 Liters Loop Fermentor/Fermentor pH (pH)/Temperature (Temp)

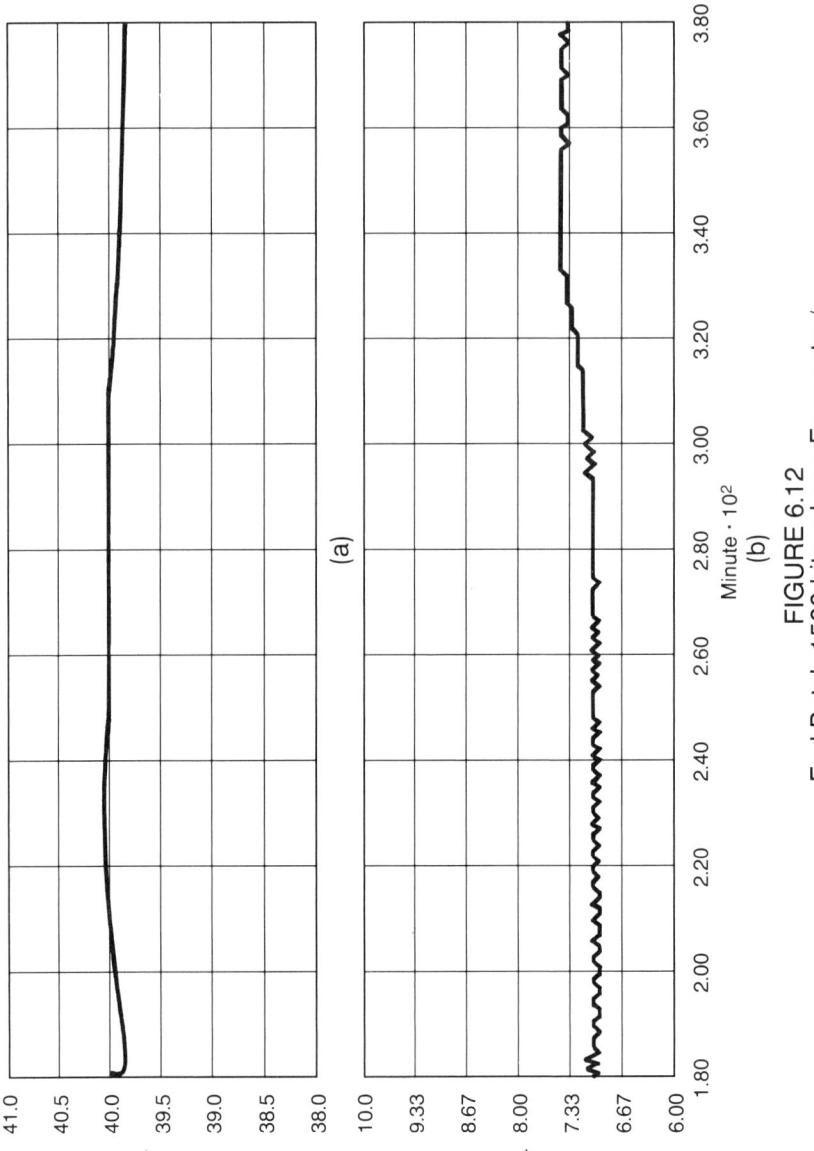

FIGURE 6.12
Fed Batch 1500 Liters Loop Fermentor/
Fermentor pH (pH)/Temperature (Temp)/Fast pH, DO, and Temperature Sensors

control dissolved oxygen at the beginning of the batch and, therefore, accept the associated cost of the additional air flow. The measurement floats far above the set point when the cell growth rate and the cell concentration are both low.

Never has a control engineer had so many choices for the manipulated variable. Fresh air flow, recycle air flow for loop fermentors, agitation rate for stirred fermentors, recycle pumping rate for loop fermentors, and operating pressure can all be manipulated to control dissolved oxygen. The manipulation of agitation is preferable to air flow because the gain is larger and the response is faster. Large top-drive fermentors have limited agitation manipulation. The manipulation of operating pressure is preferable to air flow because the gain is much more linear. Since the mass balance requires that the air flow replenish that consumed by the cells, and the agitation and operating pressure methods still require sufficient bubbles, the air flow rate must be incremented as the cells grow. The simultaneous manipulation of all three after some threshold of air flow has been established provides more bang for the bucks and operation over a smaller range. Limits must be placed on the agitation rate and operating pressure. Low agitation will cause inadequate dispersion of the reagents and nutrients. While high agitation increases the mass transfer efficiency by an increase in transfer area (bubble size), shearing may develop. The operating pressure must be high enough to keep contaminants out of the fermentor but not so high as to exceed the pressure rating of electrode assemblies or the vessel itself. For a loop fermentor, the recycle air is usually ratioed to the fresh air flow and the recycle pump speed adjusted slowly to insure oxygen starvation does not occur in the recycle line. The recycle pump speed must be high enough to provide adequate mixing in the fermentor by jet action.

Normally, pH control is very difficult because the set point is close to the equivalence point, which is the steepest point of the titration curve (point of highest gain). Biochemical processes are so well buffered that the slope of the titration curve is flat. Thus, the process gain and the open-loop error are rather small. This factor, combined with the closely regulated composition of feeds, slow load disturbances from cell growth, and low dead time-to-time constant ratio for high agitation, makes fermentor pH control relatively easy. Most of the control problems that do develop are caused by the extremely small reagent flow requirements, especially at the beginning of the fermentation. The flow rate is so small that the reagent piping and dip tube have partial flow. The reagent may just drip down the sides of the tube or fermentor wall. Also, when reagent flow stops, the submerged part of a dip tube will back-fill with fermentor broth. The result is a reagent delivery dead time (particularly when the dip tube has to be flushed out) that is much larger than any other dead time in the loop. The reagent flow requirement is so low that the flow is laminar through a control valve. Thus, a control valve especially designed for laminar flow, such as that as shown in Figure 6.1, is necessary. To date, positive displacement pumps are predominantly used for reagent delivery. On-off control of the pump is not sufficient to handle the extreme changes in reagent demand through the batch cycle. A pulse width or duration type of reagent control is used where the 'add' or 'on' time depends upon the load, and the 'mix' or 'off' time (set by the cycle time) depends upon the vessel size and agitator speed. Algorithms can be programmed in the more powerful of the microprocessor-based controllers in distributed control systems to generate a pulse width or duration that is proportional to the pH controller output by the use of timer, integrator, and signal monitor functions. For in-line systems, the pulses of reagent create considerable pH measurement noise, and the whole pH control problem becomes more difficult due to the lack of back-mixing. Since flow disturbances are too fast for such systems, the reagent flow should be ratioed to the throughput flow and the ratio corrected by the pH controller.

Temperature control is relatively easy in that the low dead time-to-time constant ratio and the use of cascade control make the control errors less than the measurement errors for a well-designed and well-installed system. The most frequent source of poor performance is too large a gap between the split-ranged steam or hot water and chilled water valves for tempered water temperature control. As fermentors move toward operating at higher and higher cell concentrations, some temperature loops run out of the manipulated variable at the end of the batch because the heat generation exceeds the capability of the heat transfer surface area.

Fermentor substrate concentration control has a low process dead time-to-time constant ratio because the main contributor to the process dead time is the turnover time, which is rather small for a well-agitated vessel. However, the sample transportation delay and analysis lag for the substrate measurement add a dead time to the loop much larger than the process dead time. Also, the secondary effects of cell growth inhibition or promotion from substrate concentration changes add some slower dynamics to the open-loop response. The result is not particularly tight control of the substrate concentration.

Fermentor cell concentration control has a large process dead time-to-time constant ratio due to the large lag in cell replication. There is also an inverse response where the cell concentration initially decreases due to dilution and then increases due to growth after an increase in substrate or nutrient flow. The controller must be tuned to ignore the initial reaction, which is the opposite of the final reaction. Sometimes the final response is so much slower than the initial response that controllers are mistakenly tuned with the wrong action (direct versus reverse) and the wrong settings (high reset action). While the short-term response looks good, the measurement will drift away from the set point as the slower dynamics take effect. The amount of inverse response depends on the relative magnitudes of growth rate and maintenance coefficients and their dependency on the substrate concentration. If the amount of substrate required for cell maintenance is compared to the amount needed for cell growth, the inverse response will be small. On top of all this, the sample transportation delay and analysis cycle time for many measurement methods are horrendous. While the response of mass spectrometers for off-gas measurement of oxygen and carbon dioxide is fast, the integration and filtering required to reduce noise in the calculated cell concentration add appreciable time delay to the loop. Some of this noise is due to interaction between variables. For example, the output plots for the stirred fermentor show a downward spike in the carbon dioxide production rate (CPR) that coincides with an upward spike in the pH of the broth. The change in pH changes the solubility of carbon dioxide in the broth, which inturn changes the amount of carbon dioxide in the off-gas even if the cell metabolic and growth rates stay constant.

An interesting control problem in the separation/purification area is in-line gradient concentration control for chromatograph elution or regeneration. The pH or conductivity of the fluid is incremented in a stepwise or continuous fashion. In one application, a microprocessor-based controller from a distributed control system was used to maximize chromatograph regeneration. The controller totalized the influent flow and computed a conductivity set point as a polynomial function of the total to provide continuous gradient concentration control. The controller also ratioed the reagent flow to the influent flow and corrected the ratio by conductivity feedback control to reduce the effect of flow disturbances to the in-line system.

CHAPTER 7

Computer Systems

It is fortunate that the strategies for biochemical control are being developed when computer systems are becoming quite common. While a computer system was once considered a luxury, it is now considered a necessity. This is largely due to the advent of the distributed control system. The hardware cost has dropped to the point where such a system is comparable in price to an analog system for medium to large size installations. More importantly, the initial and maintenance costs of software have decreased dramatically. Computer wizards speaking strange languages are no longer required. Most of the software customization is menu-driven and consists of filling in the blanks or selecting and ordering the functional blocks. Results to date show general performance benefits from better data logging and correlation, greater flexibility in designing special control strategies, more monitoring of instrument signal and control strategy integrity, better accuracy of computed coefficients, easier commissioning of special strategies by automatization, and greater operator attention to and understanding of the process via CRT graphic displays.

For biochemical pilot plants and production facilities, there are also specific benefits from the automatization of experimentation, the computation of inferred measurements, and the optimization of production. The general and specific benefits are achievable only if the computer system is properly implemented. For pharmaceutical manufacturing, the computer system must pass an audit by the Food and Drug Administration (FDA). The requirements by the FDA provide good guidelines for the application of computers to other types of biochemical production.

7.1
FDA Audits

The goal of a Food and Drug Administration (FDA) audit is to insure the system is in compliance with Current Good Manufacturing Practices (CGMP). The FDA is interested in whether the system is reliable, accurate, secure, and adequately documented. Some of the detailed items checked are as follows (Motise, 1984):

(1) What parts of the process truly depend upon computer surveillance or control?
(2) What is the exact structure of the computer system?
(3) Is the computer's physical environment (temperature, dust, humidity, and vibration) adequately protected and regulated?
(4) Is the computer's electrical environment (power fluctuations, emi, and rfi) adequately protected and regulated?
(5) Are the process control programs adequately identified and documented?
(6) Does the software protect against gross operator errors?
(7) Does the operator interface cause operator fatigue?
(8) Are essential manual backup systems easily commissioned?
(9) Have the operational limits of the system been identified and verified?
(10) Has the software been tested for the worst-case conditions?
(11) Is the software adequately protected against accidental or unauthorized revisions?
(12) Are the software records protected from damage or loss?
(13) Are process alarms properly enunciated, acknowledged, overridden, queued, and prioritized?

(14) Is a test signal from the field accurately displayed?
(15) Is a test signal displayed fast enough to recognize the dynamic response of the process?

7.2
Automated Experimentation

The biochemical reactions are so complex that most product quality or yield improvements come not by *a priori* computations and analysis but by *in situ* result tabulation and analysis. The best composition and timing of addition of the magical mixture of nutrients, substrate, and product formation initiators is arrived at only after extensive experimentation. A computer can gather data from measurements and manipulate final elements to duplicate run conditions much more accurately than a team of research technicians with clipboards. Computers don't have hangovers, emotional problems, or coffee breaks. Instead, process computers typically have 0.03% accuracy for analog input and output conversion, computational accuracy to 9 significant digits, historical trend packages for correlating data, and the ability to talk to larger computers to design experiments based on statistical analysis and linear programming techniques.

7.3
Inferred Measurements

The cell, substrate, and product concentration in a fermentor can be inferred by integration of the differential equations for the material balance of each of these components in the broth. The elemental balances from the stoichiometric equations, knowledge of the feed and outlet stream flow rates and compositions, and specific rates from the oxygen uptake rate (OUR), carbon dioxide production rate (CPR), and nitrogen uptake rate (NUR) are used to define the inputs to the differential equations. For a batch fermentation, the only outlet stream is the off-gas. A mass spectrometer (preferred for accuracy reasons) is typically used to measure the oxygen and carbon dioxide concentration in the off-gas to compute the oxygen uptake rate (OUR) and the carbon dioxide production rate (CPR). The nitrogen uptake rate (NUR) is computed as the derivative of the ammonia reagent flow, whose demand is set by pH control. The following laboratory example, from the 1984 course titled "Computer Control and Optimization of Fermentation Processes" sponsored by Rutgers University and New Brunswick Scientific®, shows how the OUR, CPR, and NUR is used to compute the cell concentration, substrate (glucose) concentration, and product concentration for an ethanol fed batch fermentation. The substrate concentration was used as the controlled variable for feedback control of the glucose feed. The results of the laboratory example, as shown in Figure 7.1, show considerable oscillations in the OUR, CPR, and NUR. The very slow frequency oscillation of the NUR follows the pH oscillation.

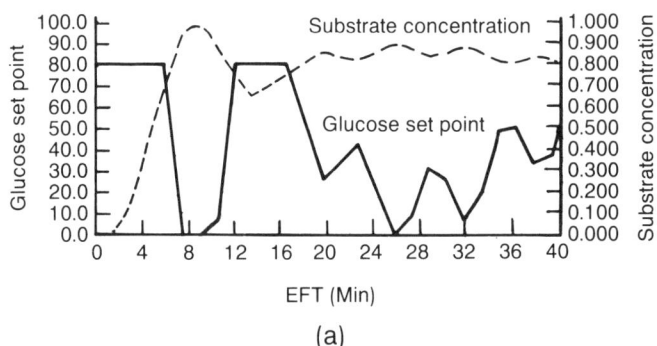

(a)

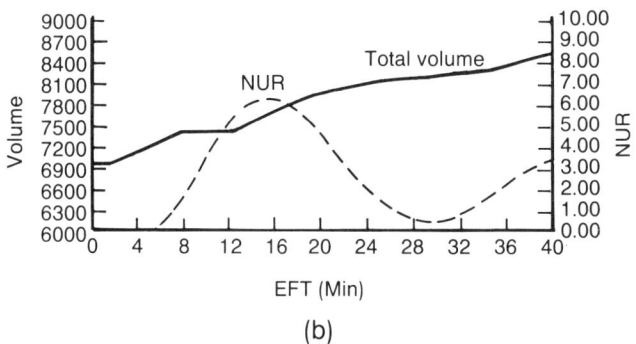

(b)

FIGURE 7.1
Laboratory Example Results for Inferred Measurements

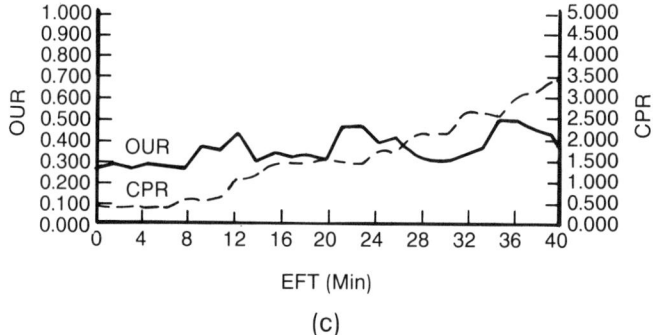

(c)

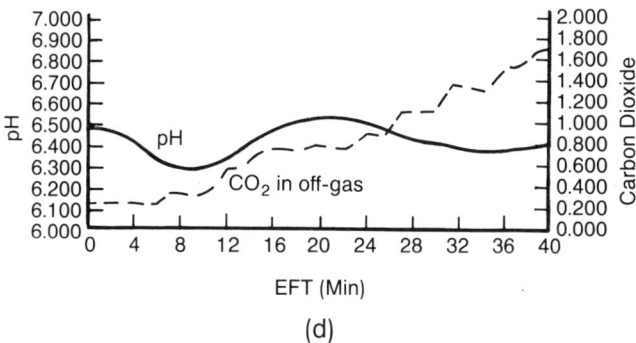

(d)

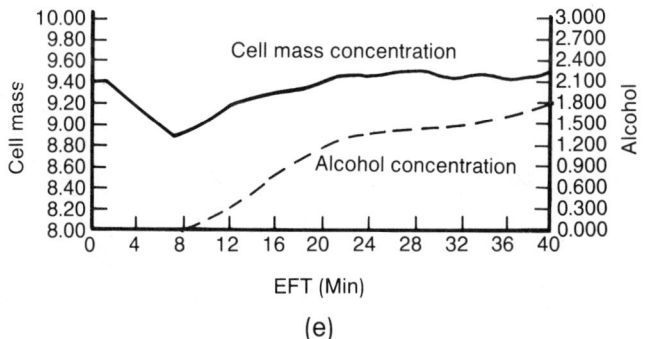

(e)

FIGURE 7.1 (continued)
Laboratory Example Results for Inferred Measurements

Material balance differential equations:

$$\frac{d(V)}{dt} = F_s + F_b \tag{7.1}$$

$$\frac{d(V \cdot X)}{dt} = u_x \cdot V \cdot X \tag{7.2}$$

$$\frac{d(V \cdot S)}{dt} = F_s \cdot S_f - u_s \cdot V \cdot X \tag{7.3}$$

$$\frac{d(V \cdot P)}{dt} = u_p \cdot V \cdot X \tag{7.4}$$

Stoichiometric equations:

$$a \cdot C_6H_{12}O_6 + b \cdot O_2 + c \cdot NH_3 = $$
$$C_gH_hO_iN_j + d \cdot CO_2 + e \cdot H_2O + f \cdot C_2H_5OH \tag{7.5}$$

Elemental balance equations:

Carbon $\rightarrow 6a = g + d + 2f$ (7.6)

Hydrogen $\rightarrow 12a + 3c = h + 2e + 6f$ (7.7)

Oxygen $\rightarrow 6a + 2b = i + 2d + e + f$ (7.8)

Nitrogen $\rightarrow c = j$ (7.9)

Solution:

$$OUR = F_a \cdot (0.20946 - O_o) \tag{7.10}$$

$$CPR = F_a \cdot (C_o - 0.00033) \tag{7.11}$$

$$NUR = d_b \cdot (F_b(t1) - F_b(t2)) / (t1 - t2) \tag{7.12}$$

$$c = j \tag{7.13}$$

$$b = c \cdot OUR/NUR \tag{7.14}$$

$$d = c \cdot \text{CPR}/\text{NUR} \tag{7.15}$$

$$f = (3c - 4b + 4d - h + 2i)/4 \tag{7.16}$$

$$a = (g + d + 2f)/6 \tag{7.17}$$

$$e = (12a + 3c - 6f - h)/2 \tag{7.18}$$

$$u_x \cdot V \cdot X = \frac{1}{c} \cdot \text{NUR} \cdot MW_x \tag{7.19}$$

$$u_s \cdot V \cdot X = \frac{a}{c} \cdot \text{NUR} \cdot MW_s \tag{7.20}$$

$$u_p \cdot V \cdot X = \frac{f}{c} \cdot \text{NUR} \cdot MW_p \tag{7.21}$$

(The material balance equations are integrated in real time after the substitution of the specific rates defined by above equations.)

where

C_o = carbon dioxide in off-gas (fractional)
d_b = density of ammonia feed (gm/liter)
F_a = air feed rate (gm/min)
F_s = substrate feed rate (liters/min)
F_b = ammonia feed rate (liters/min)
MW_x = molecular weight of cells
MW_s = molecular weight of substrate
MW_p = molecular weight of product
CPR = carbon dioxide production rate (gm/min)
OUR = oxygen uptake rate (gm/min)
NUR = nitrogen uptake rate (gm/min)
O_o = oxygen in off-gas (fractional)
P = product concentration in broth (gm/liter)
u_x = specific cell growth rate (1/min)
u_p = specific product formation rate (1/min)
u_s = specific substrate uptake rate (1/min)
S = substrate concentration in broth (gm/liter)
S_f = substrate concentration in feed (gm/liter)
$t1$ = time point for computation of ammonia flow rate derivative (min)

$t2$ = time point for computation of ammonia flow rate derivative (min)
V = volume of broth (liters)
X = cell concentration in broth (gm/liters)

7.4
Optimization

Optimization of a continuous fermentor involves maximizing a profit function that is a function of process conditions. Ordinary calculus can be used to determine the best operating point in a multidimensional space of variable process conditions. Optimization of a batch fermentor involves maximizing a profit function, which is a function of process conditions that are themselves a function of time. It is a path optimization problem that requires variational calculus for its solution.

A simpler method of optimization that is instructive involves keeping the substrate concentration constant. This strategy is particularly suited for those fermentations where the specific growth rate reaches a maximum as a function of substrate concentration. Such is the case for single-cell protein production, as shown in Figure 7.2, where the specific growth peaks at 0.2% of methanol substrate in the broth. The material balance differential equations for continuous operation simplify to the point where the cell specific growth rate can be kept constant by keeping the dilution rate (feed rate divided by volume) constant. The cell, substrate, the product concentrations all become functions of the dilution rate, as shown in Figure 7.3 for the case where there is no substrate consumption for cell maintenance and constant yield. If the maximum of the cell production rate $(D \cdot X)$ curve is close to the maximum dilution rate, a concentration controller is advised to help prevent washout. The material balance calculations for fed batch operation simplify to where the best substrate feed set point to keep the substrate concentration constant is an exponential profile, as defined by Equation 7.30 (Weigand, 1981).

For the more general case, the profile of fermentor batch temperature and pH (besides substrate feed) may need to make a function of batch time to optimize the production, since the specific growth rate and product formation rates are frequently functions of these variables. For example, optimal temperature

COMPUTER SYSTEMS   (131)

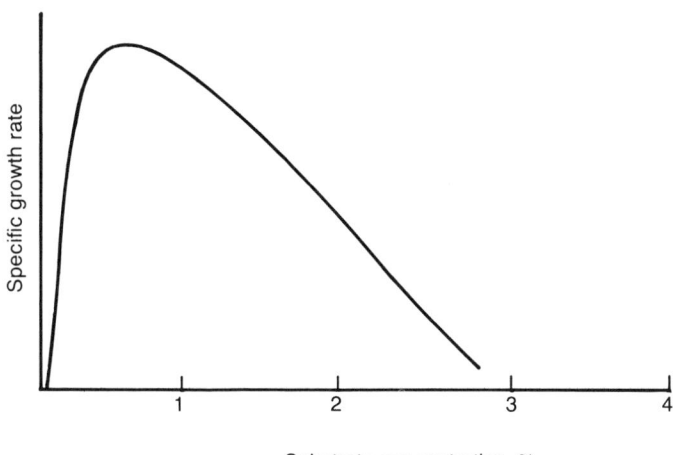

FIGURE 7.2
Specific Growth Rate as Function of Substrate Concentration

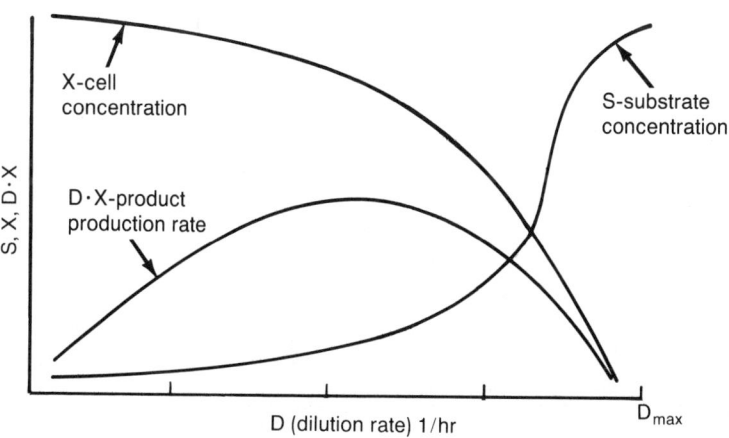

FIGURE 7.3
Yield Maximization for Continuous Fermentor

and pH profiles for gluconic acid production reduced the batch time from twelve to eight hours.

For continuous operation at steady state and constant volume:

$$\frac{dX}{dt} = R_x - \frac{F}{V} \cdot X \qquad (7.22)$$

$$\frac{dX}{dt} = 0 \qquad (7.23)$$

$$R_x = u_x \cdot X \qquad (7.24)$$

Therefore:

$$u_x = \frac{F}{V} \qquad (7.25)$$

$$u_x = D \qquad (7.26)$$

For batch operation:

$$\frac{dV}{dt} = F \qquad (7.27)$$

$$\frac{dV}{dt} = u_x \cdot V \qquad (7.28)$$

Therefore:

$$V = V_o \cdot e^{-u_x \cdot (t - t_o)} \qquad (7.29)$$

$$F = u_x \cdot V_o \cdot e^{-u_x \cdot (t - t_o)} \qquad (7.30)$$

where

F = feed rate (liters/hr)
V = broth volume (liters)
D = dilution rate (1/hr)
X = cell concentration (gm/liter)
u_x = specific growth rate (1/hr)
R_x = cell formation rate (gm/liter · hr)

References

Aharonowitz, Alan, and Cohen, Gerald, "The Microbiological Production of Pharmaceuticals," *Scientific American*, Sept. 1981, pp. 141–152.

Armiger, W. B., et al., "Analysis and Process Control of Batch Fed Production of E coli Using Culture Fluorescence," *Biotech '84*, USA, Online Publications, pp. 601–628.

Bailey, James E., and Ollis, David F., *Biochemical Engineering Fundamentals*, McGraw-Hill Book Company, 1977, pp. 312–322.

Baker, E. G., et al., "Methanol and Ammonia From Biomass," *CEP*, Dec. 1984, pp. 43–46.

Buhler, H., et al., "Redox Measurements: Principles and Problems," Ingold publication E-TH 2-1-CH, 1980, pp. 3–23.

Buhler, H., and Ingold, W., "Measuring pH and Oxygen in Fermentors," *Process Biochemistry*, Apr. 1976.

Carey, P. R., *Biochemical Applications of Raman and Resonance Raman Spectroscopies*, Academic Press, 1982, pp. 1–70.

"Measure Composition In Situ via Fiber-Optic Spectrophotometry," *Chemical Engineering*, Dec. 1985, p. 37.

Chotani, G., and Constantinides, A., "On-Line Glucose Analyzer for Fermentation Applications," *Biotechnology and Bioengineering*, vol. XXIV, 1982, pp. 2743–2745.

De Vries, Eric A., "Facts and Falacies of Vortex Flowmeters," *Hydrocarbon Processing*, Aug. 1982, pp. 75–76.

Eveleigh, Douglas E., "The Microbiological Production of Industrial Chemicals," *Scientific American*, Sept. 1981, pp. 155–177.

Fieschko, John, et al., "The Relationship between Cell Dry Weight Concentration and Culture Turbidity for a Recombinant DNA E Coli K12 Strain Producing High Levels of Human Alpha Interferon Analogue," *Biotechnology Progress*, vol. 1, no. 3, 1985, pp. 205–208.

Guilbault, George G., "Applications of Enzyme Electrodes in Analysis," *Annals New York Academy of Sciences*, 1981, pp. 285–292.

Hall, Alan, et al., "The Race for Miracle Drugs," *Business Weekly*, Jul. 22, 1985, pp. 92–96.

Humphrey, Arthur E., "Commercializing Biotechnology: Challenge to the Chemical Engineer," *CEP*, Dec. 1984, pp. 7–12.

Ingold Electrodes, Inc., "Operating Instructions for CO_2 Probe," 1982.

Kempe, Eberhard, and Schallenberger, Wolfgang, "Measuring and Control of Fermentation Processes: Part I," *Process Biochemistry*, 1983, pp. 7–12.

King, J. Derwin, "NMR for On-Stream Measurements," ISA Conference paper 85-0874, Oct. 1985, pp. 1387–1405.

Krebs, W. M., and Haddad, I. A., "The Oxygen Electrode in Fermentation Systems," reprint from volume 13 of *Developments in Industrial Microbiology*, American Institute of Biological Sciences, 1972, pp. 113–127.

Lee, Y. H., "Pulsed Light Probe for Cell Density Measurement," *Biotechnology and Bioengineering*, vol. XXIII, 1981, pp. 1903–1906.

McMillan, Gregory K., *Tuning and Control Loop Performance*, ISA, 1983, pp. 23–131, 183–211.

McMillan, Gregory K., *pH Control*, ISA, 1985, pp. 51–86.

McMillan, Gregory K., "Pressure Control: Without Dead Time, I Might be Out of a Job," *InTech*, Nov. 1985, pp. 49–53.

McMillan, Gregory K., "Advanced Control Algorithms: Beware of False Prophecies," *InTech*, Jan. 1986, pp.

Motise, Paul J., "What to Expect when FDA Audits Computer-Controlled Processes," *Pharmaceutical Manufacturing*, Jul. 1984, pp. 33–35.

Meriman, D., Van London Co., Inc., Jul. 15, 1983, memo to author.

Perkin-Elmer Manual for MGA-1200 Multiple Gas Analyzer for Industrial Applications, 1981.

Puhar, E.; Einsele, A.; Buhler, H.; and Ingold, W., "Steam Sterilizable CO_2 Electrode," *Biotechnology and Bioengineering*, vol. XXII, 1980, pp. 2411–2416.

Richmond, D. W., "Selecting Thermowells for Accuracy and Endurance," *InTech*, Feb. 1980, pp. 59–63.

Rose, Anthony H., "The Microbiological Production of Food and Drink," *Scientific American*, Sept. 1981, pp. 127–138.

Shinskey, F. G., *pH and pION Control in Process and Waste Streams*, John Wiley & Sons, 1973, pp. 20–22.

Skoog, D. A., and West, D. M., *Principle of Instrumental Analysis*, Saunders College, 1980, pp. 377–425, 477–499, 541–542.

Weigand, W. A., "Maximum Cell Productivity by Repeated Fed-Batch Culture for Constant Yield Case," *Biotechnology and Bioengineering*, vol. XXIII, 1981, pp. 249–266.

Weiss, Marvin D., "Chemically-Sensitive Transducers: A New Breed of Miniature Sensors," *InTech*, Jun. 1985, pp. 45–49.

APPENDIX

Continuous Biochemical Reactor Composition Control Example

(Source: McMillan, 1983)

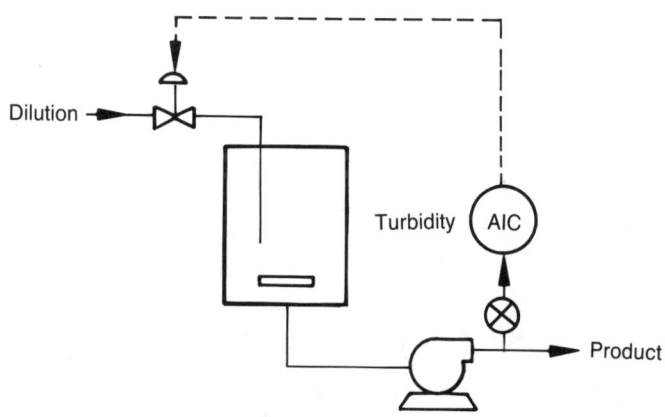

Biological Reactor Concentration Loop (Runaway)

Given:
  (a) Average nutrient concentration is 4 nanograms per milliliter (ng/ml) (C_i).
  (b) Nutrient concentration where growth rate is half its maximum is 1 ng/ml (K_i).
  (c) Maximum cell growth rate is 0.01 generation per minute (U_x).
  (d) Substrate consumption rate for cell survival is 0.001 generation per minute (K_e).
  (e) Reactor working volume is 600 gallons (V).
  (f) Average throughput flow is 10 gpm (F).
  (g) Agitator pumping rate is 200 gpm (F_a).
  (h) Loop steady-state gain is 1 ($K_v \cdot K_p \cdot K_m$).

Find:
  the proportional band window size.

Solution:
  (a) Calculate the ultimate period for the loop:

T_u = ultimate period of oscillation.
TC = time constant of the self-regulating part of the process
TD = dead time of the self-regulating part of the process
TC' = time constant of the positive feedback part of the process

$$T_u = 4 \cdot \left[1 + \left[\frac{N}{D}\right]^{0.65}\right] \cdot TD$$

$$N = (TC' + TC) \cdot TC' \cdot TC$$

$$D = (TC' - TC) \cdot (TC' - TD) \cdot TD$$

$$TD = \frac{V}{F_a + F} = \frac{600}{200 = 10} = 3 \text{ minutes}$$

$$TC = \frac{V}{F} - TD = \frac{600}{10} - 3 = 57 \text{ minutes}$$

$$TC' = 1/U$$

$$U = U_x \cdot \left[\frac{C_i}{K_i + C_i}\right] - K_e = 0.01 \cdot \left[\frac{4}{1 + 4}\right]$$
$$- 0.001 = 0.007$$

APPENDIX A

$$TC' = 1/0.007 = 142 \text{ minutes}$$
$$N = (142 + 57) \cdot 142 \cdot 57 = 1{,}610{,}706$$
$$D = (142 - 57) \cdot (142 - 3) \cdot 3 = 35{,}445$$
$$T_u = 4 \cdot \left[1 + \left[\frac{1{,}610{,}706}{35{,}445} \right]^{0.65} \right] \cdot 3 = 155 \text{ minutes}$$

(b) Calculate the minimum and maximum proportional bands:

$$PB_{max} = 80 \cdot K_v \cdot K_p \cdot K_m = 80 \text{ percent}$$
$$PB_{min} = 0.5 \cdot PB$$
$$PB = \frac{K_u \cdot 100 \cdot T_u^2 \cdot K_v \cdot K_p \cdot K_m}{(2\pi)^2 \cdot TC \cdot TC'} = \frac{3.603.750}{319.214}$$
$$= 11 \text{ percent}$$
$$PB_{min} = 0.5 \cdot 11 = 5.5 \text{ percent}$$

(c) Calculate the PB window size:

$$\frac{PB_{max}}{PB_{min}} = \frac{80}{5.5} = 14.5$$

Conclusions:

The proportional band window size is slightly less than the minimum window size of 15 recommended by Luyben. An increase in nutrient concentration will increase the growth rate, which will decrease the runaway time constant, which will decrease the window size. However, the growth rate is already close to its maximum. Any analysis measurement dead time for nutrient dilution control or substrate activation delays will decrease the window size. The ultimate period is so large that tuning will be tedious. Also, the integral and derivative time settings required are beyond the range of most conventional analog controllers. If the integral time setting used is too small, the loop will go into a slow reset cycle.

New Information:
Turbidity meter sample transportation delay is 2 minutes.

Find:
the new PB window size.

Solution:

(a) Calculate the new ultimate period for the analysis dead time:

$$T_u = 4 \cdot \left[1 + \left[\frac{N}{D} \right]^{0.65} \right] \cdot TD$$

$N = (TC' + TC) \cdot TC' \cdot TC = 1{,}610{,}706$

$D = (TC' - TC) \cdot (TC' - TD) \cdot TD$

$D = (142 - 57) \cdot (142 - 5) \cdot 5 = 58{,}225$

$$T_u = 4 \cdot \left[1 \div \left[\frac{1{,}610{,}706}{58{,}225} \right]^{0.65} \right] \cdot 5 = 193 \text{ minutes}$$

(b) Calculate the minimum and maximum proportional bands:

$PB_{max} = 80 \cdot K_v \cdot K_p \cdot K_m = 80$ percent

$PB_{min} = 0.5 \cdot PB$

$$PB = \frac{K_u \cdot 100 \cdot Tu^2 \cdot K_v \cdot K_p \cdot K_m}{(2\pi)^2 \cdot TC \cdot TC'} = \frac{5{,}592{,}504}{319{,}214}$$

$= 18$ percent

$PB_{min} = 0.5 \cdot 18 = 9$ percent

(c) Calculate the *PB* window size:

$$\frac{PB_{max}}{PB_{min}} = \frac{80}{9} = 8.9$$

Conclusions:

The window size decreased significantly. Sample transportation delays must be minimized.

APPENDIX

ACSL Listing for Dynamic Simulation of Fermentor

```
PROGRAM BATCH OR CONTINUOUS FERMENTOR CONTROL
'PROGRAM MASS UNITS ARE GRAMS'
'PROGRAM LENGTH UNITS ARE CENTIMETERS'
'PROGRAM FORCE UNITS ARE DYNES'
'PROGRAM VISCOSITY UNITS ARE CENTIPOISE'
'PROGRAM PRESSURE UNITS ARE MILLIMETERS OF MERCURY ABSOLUTE'
'PROGRAM TEMPERATURE UNITS ARE DEGREES CENTIGRADE'
'PROGRAM TIME UNITS ARE MINUTES (EXCEPT DIFFUSIVITIES USE SECONDS)'
'PROGRAM GROWTH RATES ARE SPECIFIC GROWTH RATES'
'INPUT PARAMETERS:'
'CC   = HEAT CAPACITY OF COOLANT (CAL/GM C)'
'CG   = HEAT CAPACITY OF GAS FEED (CAL/GM C)'
'CL   = HEAT CAPACITY OF LIQ DILUTION (CAL/GM C)'
'CS   = HEAT CAPACITY OF SUBSTRATE FEED (CAL/GM C)'
'CW   = HEAT CAPACITY OF COIL WALL (CAL/GM C)'
'DC   = DENSITY OF COOLANT (GM/CU CM)'
'DG   = DENSITY OF GAS (GM/CU CM)'
'DW   = DENSITY OF COIL WALL (GM/CU CM)'
'DL   = DENSITY OF LIQ DILUTION (GM/CU CM)'
'DS   = DENSITY OF SUBSTRATE FEED (GM/CU CM)'
'FFG  = FEEDFORWARD GAS RAMP RATE FOR FED BATCH MODE (1/MIN)'
'FFS  = FEEDFORWARD SUBSTRATE RAMP RATE FOR FED BATCH MODE (1/MIN)'
```

(141)

APPENDIX B

```
'G0   = PRODUCT GROWTH RATE INDEPENDENT OF CELL GROWTH RATE (1/MIN)'
'G1   = PRODUCT GROWTH RATE FRACTION OF CELL GROWTH RATE'
'HN   = THE NUMBER OF HYDROGEN OR HYDROXL IONS ARRAY FOR PHION MACRO'
'HS   = HEAT OF COMBUSTION OF SUBSTRATE (CAL/GM)'
'HV   = HEAT OF VAPORIZATION OF WATER (CAL/GM)'
'HX   = HEAT OF COMBUSTION OF CELLS (CAL/GM)'
'KM   = SUBSTRATE CONCENTRATION FOR HALF MAX GROWTH RATE (GM/CU CM)'
'MC   = MAINTENANCE COEFFICIENT FOR CARBON DIOXIDE (GM CO2/GM CELL/MIN)'
'MO   = MAINTENANCE COEFFICIENT FOR OXYGEN (GM O2/GM CELL/MIN)'
'MS   = MAINTENANCE COEFFICIENT FOR SUBSTRATE (GM SUBSTRATE/GM CELL/MIN)'
'MW   = MOLECULAR WEIGHT ARRAY FOR PHION MACRO'
'MWG  = MOLECULAR WEIGHT OF GAS FEED'
'MWF  = MOLECULAR WEIGHT OF LIQ CONTENTS OF FERMENTOR'
'PC   = PARTIAL PRESSURE OF CARBON DIOXIDE IN GAS FEED (MM HG)'
'PHL  = LOWEST PH THAT DOES NOT AFFECT MAX GROWTH RATE (PH)'
'PHH  = HIGHEST PH THAT DOES NOT AFFECT MAX GROWTH RATE (PH)'
'PT   = PRESSURE (TOTAL) OF GAS FEED (MM HG)'
'PO   = PARTIAL PRESSURE OF OXYGEN IN GAS FEED (MM HG)'
'PP   = PARAMETER ARRAY FOR PHION MACRO'
'PK1  = FIRST DISSOCIATION CONSTANT ARRAY FOR PHION MACRO'
'PK2  = SECOND DISSOCIATION CONSTANT ARRAY FOR PHION MACRO'
'RF   = RADIUS OF FERMENTOR (CM)'
'RI   = RADIUS OF IMPELLER (CM)'
'RG   = GAS LAW CONSTANT (MM HG CU CM/GM MOLE K)'
'RPMIC= INITIAL SPEED OF AGITATOR (RPM)'
'SC   = SOLUBILITY OF CARBON DIOXIDE IN LIQUID (CU CM/CU CM)'
'SO   = SOLUBILITY OF OXYGEN IN LIQUID (CU CM/CU CM)'
'ST   = SURFACE TENSION (DYNES/CM)'
'TG   = TEMPERATURE OF GAS FLOW (DEG C)'
'TL   = TEMPERATURE OF LIQ DILUTION (DEG C)'
'TS   = TEMPERATURE OF SUBSTRATE FEED (DEG C)'
'UM   = VISCOSITY OF FERMENTOR BROTH AT KM (CP)'
'VC   = VOLUME OF COIL (CU CM)'
'VW   = VOLUME OF COIL WALL (CU CM)'
'YPC  = PRODUCT-CARBON DIOXIDE YIELD (GM PRODUCT/GM CO2)'
'YPO  = PRODUCT-OXYGEN YIELD (GM PRODUCT/GM O2)'
'YPS  = PRODUCT-SUBSTRATE YIELD (GM PRODUCT/GM SUBSTRATE)'
'YXC  = CELL-CARBON DIOXIDE YIELD (GM CELLS/GM CO2)'
'YXO  = CELL-OXYGEN YIELD (GM CELLS/GM O2)'
'YXS  = CELL-SUBSTRATE YIELD (GM CELLS/GM SUBSTRATE)'
'ZO   = DIFFUSIVITY OF OXYGEN (SQ CM/SEC)'
'ZC   = DIFFUSIVITY OF CARBON DIOXIDE (SQ CM/SEC)'
'CALCULATED PARAMETERS:'
'AE   = ARRHENIUS EQUATION CORRECTION OF MAINTENANCE COEFFICIENT'
'AF   = CROSS SECTIONAL AREA OF FERMENTOR (SQ CM)'
'AI   = INTERFACIAL AREA PER UNIT VOLUME (1/CM)'
'BD   = BUBBLE DIAMETER (CM)'
'BT   = BUBBLE RISE TIME (MIN)'
'BV   = BUBBLE VELOCITY (CM/MIN)'
'COGO = CONCENTRATION OF OXYGEN IN OFFGAS (GM/CU CM)'
'COCO = CONCENTRATION OF CARBON DIOXIDE IN OFFGAS (GM/CU CM)'
'CWGI = CONCENTRATION OF WATER IN GAS FEED (GM/CU CM)'
'CWGO = CONCENTRATION OF WATER IN OFFGAS (GM/CU CM)'
'CF   = HEAT CAPACITY OF FERMENTOR LIQUID CONTENTS (CAL/GM C)'
'DF   = DENSITY OF FERMENTOR LIQUID CONTENTS (GM/CU CM)'
'DPF  = DIFFERENTIAL PRESSURE DRIVING FORCE FOR MASS TRANSFER (CM/SEC)'
'FA   = AGITATOR PUMPING RATE (CU CM/MIN)'
'FAIMAX = MAXIMUM AMMONIA FEED FLOW (CU CM/MIN)'
'FCIMAX = MAXIMUM COOLANT FEED FLOW (CU CM/MIN)'
'FGIMAX = MAXIMUM GAS FEED FLOW (CU CM/MIN)'
'FLOMAX = MAXIMUM DILUTION FLOW (CONTINUOUS MODE) (CU CM/MIN)'
'FSIMAX = MAXIMUM SUBSTRATE FEED FLOW (CU CM/MIN)'
```

APPENDIX B

```
'GP    = GROWTH RATE OF PRODUCT (1/MIN)'
'GX    = GROWTH RATE OF CELL MASS (1/MIN)'
'HC    = MOLE DISTRIBUTION COEFF FOR CARBON DIOXIDE FROM HENRYS LAW'
'HF    = HEIGTH OF LIQUID IN FERMENTOR (CM)'
'HO    = MOLE DISTRIBUTION COEFF FOR OXYGEN FROM HENRYS LAW'
'HP    = HORSEPOWER OF AGITATION (HP)'
'JC    = MASS TRANSFER RATE OF CARBON DIOXIDE (GM/MIN)'
'JO    = MASS TRANSFER RATE OF OXYGEN (GM/MIN)'
'KC    = MASS TRANSFER COEFF OF CARBON DIOXIDE (GM/MIN/SQ CM)'
'KO    = MASS TRANSFER COEFF OF OXYGEN (GM/MIN/SQ CM)'
'NQ    = IMPELLER DISCHARGE COEFFICIENT'
'NSC   = SCHMIDT NUMBER FOR CARBON DIOXIDE'
'NSO   = SCHMIDT NUMBER FOR OXYGEN'
'PH    = PH OF FERMENTOR (PH)'
'QA    = HEAT GENERATION DUE TO AGITATION (CAL/MIN)'
'QG    = HEAT GENERATION DUE TO CELL GROWTH (CAL/MIN)'
'QS    = SENSIBLE HEAT GAIN OR LOSS (CAL/MIN)'
'QV    = HEAT OF EVAPORATION LOSS (CAL/MIN)'
'RC    = RATE OF CARBON DIOXIDE FORMATION (GM/CU CM/MIN)'
'RN    = REYNOLDS NUMBER DUE TO AGITATION'
'RO    = RATE OF OXYGEN CONSUMPTION (GM/CU CM/MIN)'
'RP    = RATE OF PRODUCT FORMATION (GM/CU CM/MIN)'
'RS    = RATE OF SUBSTRATE CONSUMPTION (GM/CU CM/MIN)'
'TD    = TIME LAG FROM MIXING (MIN)'
'UF    = VISCOSITY OF FERMENTOR LIQUID CONTENTS (GM/CM/SEC)'
'UMAX  = MAXIMUM GROWTH RATE (1/MIN)'
'VG    = GAS VOLUME IN FERMENTOR (CU CM)'
'VL    = LIQ VOLUME IN FERMENTOR (CU CM)'
'VF    = GAS PLUS LIQ VOLUME IN FERMENTOR (CU CM)'
'WP    = THE WEIGHT PER CENT ARRAY FOR PHION MACRO'
'XC    = MOLE FRACTION OF CARBON DIOXIDE IN FERMENTOR LIQUID'
'XO    = MOLE FRACTION OF OXYGEN IN FERMENTOR LIQUID'
'YCE   = GAS MOLE FRACTION EXPONENTIAL PROFILE FACTOR FOR CARBON DIOXIDE'
'YCI   = MOLE FRACTION OF CARBON DIOXIDE IN FEED GAS'
'YCO   = MOLE FRACTION OF CARBON DIOXIDE IN OFFGAS'
'YOE   = GAS MOLE FRACTION EXPONENTIAL PROFILE FACTOR FOR OXYGEN'
'YOI   = MOLE FRACTION OF OXYGEN IN FEED GAS'
'YOO   = MOLE FRACTION OF OXYGEN IN OFFGAS'
'OUTPUT VARIABLES FOR PLOTS'
'GAS   = GAS FLOW RATE (CU CM/MIN)'
'CELL  = CELL MASS CONCENTRATION IN FERMENTOR (GM/CU CM)'
'CO2   = CARBON DIOXIDE WEIGHT FRACTION IN OFFGAS'
'CPR   = CARBON DIOXIDE PODUCTION RATE (GM/MIN)'
'DILUTE= DILUTION FLOW RATE (CU CM/MIN)'
'FEED  = SUBSTRATE FEED RATE (CU CM/MIN)'
'FOOD  = SUBSTRATE CONCENTRATION IN FERMENTOR (GM/CU CM)'
'GROWTH= SPECIFIC GROWTH RATE (1/MIN)'
'OUR   = OXYGEN UPTAKE RATE (GM/MIN)'
'O2    = OXYGEN WEIGHT FRACTION IN OFFGAS'
'DO    = DISSOLVED OXYGEN CONCENTRATION IN FERMENTOR (PPM)'
'PROD  = PRODUCT CONCENTRATION IN FERMENTOR (GM/CU CM)'
'TEMP  = TEMPERATURE OF FERMENTOR (DEG C)'
'INPUT DATA TABLES:'
'REMOTE = CELL MASS REMOTE SETPOINT VS BATCH TIME'
'GROW   = MAX GROWTH RATE VS FERMENTOR TEMP'
'CARBON = HENRYS LAW CONSTANT FOR CARBON DIOXIDE VS FERMENTOR TEMP'
'POWER  = POWER NUMBER VS AGITATION REYNOLDS NUMBER'
'OXYGEN = HENRYS LAW CONSTANT FOR OXYGEN VS FERMENTOR TEMP'
'INTEGRATOR OUTPUTS (AN INITIAL CONDITION IS NEEDED):'
'CALO  = CONCENTRATION OF AMMONIA REAGENT IN FERMENTOR LIQ (GM/CU CM)'
'CCLO  = CONCENTRATION OF CARBON DIOXIDE IN FERMENTOR LIQ (GM/CU CM)'
'COLO  = CONCENTRATION OF OXYGEN IN FERMENTOR LIQ (GM/CU CM)'
```

```
'CPLO = CONCENTRATION OF PRODUCT IN FERMENTOR LIQ (GM/CU CM)'
'CSLO = CONCENTRATION OF SUBSTRATE IN FERMENTOR LIQ (GM/CU CM)'
'CXLO = CONCENTRATION OF CELLS IN FERMENTOR LIQ (GM/CU CM)'
'TC   = TEMPERATURE OF COOLANT ARRAY (DEG C)'
'TF   = TEMPERATURE OF FERMENTOR LIQ CONTENTS (DEG C)'
'TW   = TEMPERATURE OF COIL WALL ARRAY (DEG C)'
'TIME LAGGED OUTPUTS (AN INITIAL CONDITION AND TIME LAG IS NEEDED):'
'CXM  = CELL MASS CONCENTRATION MEASUREMENT (PPM)'
'DOM  = DISSOLVED OXYGEN MEASUREMENT (PPM)'
'PHM  = PH MEASUREMENT (PH)'
'TFM  = TEMPERATURE MEASUREMENT (PH)'
INITIAL
INTEGER I,N
ARRAY PP(6),CI(3),WP(3),MW(3),PK1(3),PK2(3),HN(3)
ARRAY TC(5),TW(5),QW(5),QF(5)
ARRAY TCDT(5),TWDT(5),TCIC(5),TWIC(5)
ARRAY CXC(12),DOC(12),PHC(12),TFC(12)
'DEFINE THE MACRO FOR HEAT TRANSFER TO THE COIL IN THE FERMENTOR'
MACRO COIL(TCDT,TWDT,QC,QW,QF,TCO,TCI,FC,DC,CC,VC,...
N,UA,TC,TW,TF,FCNORM,DW,CW,VW)
PROCEDURAL(TCDT,TWDT,QC,QW,QF,TCO=TCI,FC,DC,CC,VC,...
N,UA,TC,TW,TF,FCNORM,DW,CW,VW)
MACRO REDEFINE I,J,M
INTEGER I,J,M
NN=N
J=N-1
QW(1)=UA/NN*(TC(1)-TW(1))*(FC/FCNORM)**0.8
QF(N)=UA/NN*(TW(N)-TF)
TCDT(1)=((TCI  -TC(1+1))/2.*FC*CC-QW(1))/(DC*VC/NN*CC)
DO LOOP1 I=2,J
M=N-I+1
QW(I)=UA/NN*(TC(I)-TW(I))*(FC/FCNORM)**0.8
QF(M)=UA/NN*(TW(M)-TF)
TCDT(I)=((TC(I-1)-TC(I+1))/2.*FC*CC-QW(I))/(DC*VC/NN*CC)
LOOP1.. CONTINUE
QW(N)=UA/NN*(TC(N)-TW(N))*(FC/FCNORM)**0.8
QF(1)=UA/NN*(TW(1)-TF)
TCDT(N)=((TC(J)-TCO)/2.*FC*CC-QW(N))/(DC*VC/NN*CC)
TCO=TC(N)+TC(N)-TC(J)
QC=0.
DO LOOP2 I=1,N
TWDT(I)=(-QF(I)+QW(I))/(DW*VW/NN*CW)
QC=-QF(I)+QC
LOOP2.. CONTINUE
END
MACRO END
'SEE THE ACSL FUNCTIONAL BLOCK DOCUMENTATION FOR CNTRL AND PHION MACROS'
'SET THE TABLE INPUT DATA'
TABLE GROW,1,7/...
10.,20.,30.,38.,40.,42.,47.,...
-7.9,-6.3,-4.9,-4.2,-4.1,-4.2,-7.9/
TABLE OXYGEN,1,7/...
10.,20.,30.,35.,40.,45.,50.,...
2.49E7,3.04E7,3.61E7,3.85E7,4.06E7,4.28E7,4.47E7/
TABLE CARBON,1,7/...
10.,20.,30.,35.,40.,45.,50.,...
0.79E6,1.07E6,1.41E6,1.58E6,1.77E6,1.95E6,2.15E6/
TABLE REMOTE,1,10/...
0.,60.,120.,180.,240.,300.,360.,420.,480.,540.,...
0.5E-5,0.1E-4,0.5E-4,2.4E-4,6.0E-4,22.E-4,33.E-4,48.E-4,50.E-4,52.E-4/
TABLE POWER,1,6/...
1.,10.,100.,1000.,10000.,100000.,...
```

APPENDIX B

```
      70.,8.,4.,5.,6.,7./
'SET THE PH MACRO INPUT PARAMETERS'
CONSTANT PP=13.26,0.,1.,2.,12.,0.05
CONSTANT MW=17.03,44.01,82.
CONSTANT HN=1.,-2.,-2.
CONSTANT PK1=9.24,6.35,2.00
CONSTANT PK2=20.0,10.3,6.40
'SET THE LOWER AND UPPER LIMITS OF NO PH EFFECT ON MAX GROWTH RATE'
CONSTANT PHL=6.,PHH=8.
'SET THE INSTRUMENT LAG TIMES'
CONSTANT CXMTC=1.0,DOMTC=0.5,PHMTC=0.5,TFMTC=0.5
'SET THE INITIAL CONTROLLER OUTPUTS AND VALVE POSITIONS'
'CONTROL VALVE RANGEABILITY OF 10000:1 NEEDED TO CONTROL AT TIME=0'
CONSTANT CXCOIC=0.44,DOCOIC=0.001,PHCOIC=0.01,TFCOIC=0.01
'SET THE MAXIMUM FLOW RATE AND FEEDFORWARD PARAMETERS'
CONSTANT FSIMAX=2.0E2,FFS=0.0,FFG=0.0,FGIMAX=1.0E6
'SET THE CONCENTRATION AND VISCOSITY AT HALF MAX GROWTH RATE'
CONSTANT KM=1.5E-3,UM=1.
'SET THE SURFACE TENSION AND DIFFUSIVITIES'
CONSTANT ST=68.45,ZO=2.4E-5,ZC=2.E-5
'SET THE PRODUCT GROWTH RATE PARAMETERS'
CONSTANT GO=0.,G1=1.
'SET THE FERMENTOR VESSEL AND AGITATOR PARAMETERS'
CONSTANT RF=45.,RI=15.,RPMIC=100.,VFIC=1.0E6
'SET THE SUBSTRATE FLOW PARAMETERS'
CONSTANT DS=1.16,CS=0.39,TS=25.
'SET THE DILUTION FLOW PARAMETERS'
CONSTANT DL=1.00,CL=0.39,TL=25.
'SET THE GAS FLOW PARAMETERS'
CONSTANT DG=0.0012,CG=0.24,TG=25.,MWG=28.
'SET THE MAXIMUM LIQUID CONCENTRATIONS OF OXYGEN AND CARBON DIOXIDE'
CONSTANT SO=0.023,SC=0.530
'SET THE MAXIMUM MASS TRANSFER RATES'
CONSTANT JOMAX=0.15,JCMAX=-3.2E-04
'SET THE COIL COOLANT PARAMETERS'
CONSTANT DC=1.00,CC=1.00,TCI=10.,VC=10000.,TCO=20.
'SET THE COIL WALL PARAMETERS'
CONSTANT DW=7.82,CW=0.12,VW=1000.,N=5
'SET THE HEAT LOSS TO THE JACKET WALLS'
CONSTANT QJ=0.
'SET THE PARTIAL PRESSURES, TOTAL PRESSURE, AND GAS LAW CONSTANT'
CONSTANT PW=45.6,PO=319.,PC=0.38,PT=1520.,RG=62361.
'SET THE MAXIMUM YIELD PARAMETERS'
CONSTANT YXS=5.40E-1,YXO=1.14   ,YXC=1.23
CONSTANT YPS=5.40E-2,YPO=1.14E-1,YPC=1.23E-1
'SET THE INITIAL SUBSTRATE CONCENTRATION'
CONSTANT CSLOIC=3.5E-3
'SET THE AMMONIA AND SUBSTRATE FEED CONCENTRATIONS'
CONSTANT CALI=1.,CSLI=1.
'SET THE MAINTENANCE COEFFICIENTS'
CONSTANT MS=1.17E-3,MO=8.21E-4,MC=9.68E-4
'SET THE HEAT OF COMBUSTION AND VAPORIZATION PARAMETERS'
CONSTANT HS=4400.,HX=5800.,HV=540.
'SET THE MOLECULAR WEIGHT OF THE LIQ CONTENTS OF THE FERMENTOR'
CONSTANT MWF=18.
'SET THE CONTROLLER SETPOINTS'
CONSTANT DOSP=10.,PHSP=7.0,TFSP=40.
'SET THE CONTROLLER MODE PARAMETERS'
'THE CONCENTRATION CONTROL ACTION IS OPPOSITE FOR CONTINUOUS DILUTION'
CONSTANT CXC= -100.,.1,0.,0.,3.,0.,1.E-2,0.,1.,1.,0.,1.E-2
CONSTANT DOC= -200.,.2,1.,1.,2.,0.,30.,0.,1.,1.,0.,30.
CONSTANT PHC= -100.,.2,0.,1.,3.,2.,12.,0.,1.,1.,2.,12.
```

```
CONSTANT TFC= +50.,.2,1.,1.,3.,0.,50.,0.,1.,1.,0.,50.
'SET THE CONTROL MODE OPTION (CCC=1.-AIR AND CCC=0.-SPEED)'
CONSTANT CCC=1.
'SET THE OPERATION MODE OPTION (OOO=1.-BATCH AND OOO=0.-CONTINUOUS)'
CONSTANT OOO=1.
'SET THE ANALYSIS MODE OPTION (AAA=1.-OXYGEN AND AAA=0.-CARBON DIOXIDE)'
CONSTANT AAA=1.
'SET THE BATCH START TIME'
VARIABLE MINUTE=180.
CXSP=REMOTE(180.)
'CALCULATE THE MAXIMUM SPEED REQUIREMENT'
RPMMAX=5.*RPMIC
'CALCULATE THE WATER CONTENT OF THE GAS'
CWGI=0.
CWGO=(PW/PT)*(18./28.)*DG
'CALCULATE THE INITIAL FERMENTOR LIQUID VISCOSITY AND DENSITY'
UFIC=BOUND(0.01,10.,(UM*0.01)*(CXSP/KM))
DFIC=(1. + CXSP + CSLOIC)
PP(3)=DFIC
'CALCULATE THE GAS FEED CONCENTRATIONS'
CCGI=(PC/PT)*(44./28.)*DG
COGI=(PO/PT)*(32./28.)*DG
'CALCULATE THE INITIAL CONDITIONS BASED ON THE SET POINTS'
PHIC=PHSP
CXMIC=CXSP
DOMIC=DOSP
PHMIC=PHSP
TFMIC=TFSP
TFIC=TFSP
CXLOIC=CXSP
CXOMIC=CXSP
CXCMIC=CXSP
CPLOIC=CXSP*0.1
COLOIC=DOSP*DFIC/1.E6
COLMAX=SO*32./MWG*DG/DFIC
CCLMAX=SC*44./MWG*DG/DFIC
UMAXIC=(2.718**GROW(TFIC))/(10.**ABS(DEAD(PHL-7.,PHH-7.,PHIC-7.)))
GXIC=UMAXIC*(CSLOIC/(KM + CSLOIC))
GPIC=G1*GXIC + GO
RPIC=GPIC*CPLOIC
RXIC=GXIC*CXLOIC
RSIC=(GXIC*CXLOIC/YXS + GPIC*CPLOIC/YPS + MS*CXLOIC)
ROIC=(GXIC*CXLOIC/YXO + GPIC*CPLOIC/YPO + MO*CXLOIC)
RCIC=(GXIC*CXLOIC/YXC + GPIC*CPLOIC/YPC + MC*CXLOIC)
CCLOIC=(RCIC/ROIC)*(COLOIC/(10.**(PHSP-6.317) + 1.))
CONSTANT WP(3)=1.E-6
WP(2)=CCLOIC/DFIC
WP(1)=0.2*WP(2)
LOOP4.. CONTINUE
WP(1)=WP(1) + 0.01*WP(2)
PHION(PHIC,CI=PP,WP,MW,HN,PK1,PK2,3)
IF (PHIC.LT.PHSP) GO TO LOOP4
CALOIC=WP(1)*DFIC
'SET THE INITIAL COOLANT AND WALL TEMPERATURE PROFILES FOR THE COIL'
X=(TCI - TFIC)/(TCO - TFIC)
NN=N+1
DO LOOP3 I=1,N
II=I
TCIC(I)=(TCI - TFIC)/(2.718**(II/NN*ALOG(X))) + TFIC
TWIC(I)=(TCIC(I) - TFIC)/2. + TFIC
LOOP3.. CONTINUE
'CALCULATE THE OVERALL HEAT TRANSFER COEFFICIENT AND AREA REQUIRED'
```

APPENDIX B

```
QGIC=(GXIC*CXLOIC*VFIC)*(HS - YXS*HX)/YXS
RNIC=10.754*(RPMMAX/5.)*((RI/1.27)**2)*DFIC/UFIC
QAIC=641616./60.*POWER(RNIC)*(RPMIC**3)*((RI/1.27)**5)*DFIC/1.523E13
QSIC=DOCOIC*FGIMAX*DG*CG*(TG - TFIC)
QVIC=HV*(CWGO - CWGI)*DOCOIC*FGIMAX
QDIC=QGIC + QAIC + QSIC - QVIC
UA=QDIC/ALOG(X)
'CALCULATE THE MAXIMUM FLOW REQUIREMENTS'
FCIMAX=QDIC/(DC*CC*(TCO - TCI))/TFCOIC
FAIMAX=(WP(1)/WP(2))*RCIC*VFIC/PHCOIC
FLOMAX=RXIC*VFIC/CXCOIC
'SET THE NORMAL FLOW FOR THE HEAT TRANSFER COEFFICIENT'
FCNORM=TFCOIC*FCIMAX
'CALCULATE THE CROSS SECTIONAL AREA AND IMPELLER DISCHARGE COEFF'
AF=3.14*RF**2.
NQ=0.4/((RI/RF)**0.55)
'CALCULATE THE INITIAL GAS HOLDUP VOLUME'
FAIC=28312.*NQ*RPMIC*(RI/1.27/12.)**3
HFIC=VFIC/AF
BVIC=27.43*60. + 2.*DOCOIC*FGIMAX/AF + FAIC/AF
BTIC=HFIC/BVIC
VGIC=DOCOIC*FGIMAX*BTIC
'CALCULATE THE MIXING TIME LAG'
MTC=VFIC/FAIC
'SET THE SIMULATION FINISH TIME, TIME VARIABLE, AND TIME INTERVAL'
CONSTANT STOP=600.
ALGORITHM IALG=1
CINTERVAL CINT=1.
END
DYNAMIC
DERIVATIVE
'SIMULATE THE MATERIAL BALANCE DYNAMICS FOR THE LIQUID PHASE'
CPLO=BOUND(0.,1.,INTEG(REALPL(MTC,(     RP*VL - CPLO*FLO)/VL,...
0.),CPLOIC))
CXLO=BOUND(0.,1.,INTEG(REALPL(MTC,(     RX*VL - CXLO*FLO)/VL,...
0.),CXLOIC))
CALO=BOUND(0.,1.,INTEG(REALPL(MTC,( CALI*FAI - CALO*FLO)/VL,...
0.),CALOIC))
CSLO=BOUND(0.,1.,INTEG(REALPL(MTC,( CSLI*FSI - CSLO*FLO - RS*VL)/VL,...
0.),CSLOIC))
COLO=BOUND(0.,COLMAX,INTEG(REALPL(MTC,(   JO  - COLO*FLO - RO*VL)/VL,...
0.),COLOIC))
CCLO=BOUND(0.,CCLMAX,INTEG(REALPL(MTC,(RC*VL - CCLO*FLO - JC   )/VL,...
0.),CCLOIC))
'SIMULATE THE MATERIAL BALANCE DYNAMICS FOR THE GAS PHASE'
HO=OXYGEN(TF)/PT
HC=CARBON(TF)/PT
YOE=BOUND(-10.,0.,(-6*KO*BT*MWG)/(BD*DG*HO))
YCE=BOUND(-10.,0.,(-6*KC*BT*MWG)/(BD*DG*HC))
YOO=HO*XO + (YOI - HO*XO)*2.718**YOE
YCO=HC*XC + (YCI - HC*XC)*2.718**YCE
YOI=COGI/DG*MWG/32.
YCI=CCGI/DG*MWG/44.
XO=COLO/DF*MWF/32.
XC=CCXO/DF*MWF/44.
COGO=YOO*DG*32./MWG
CCGO=YCO*DG*44./MWG
'CALCULATE THE REACTION RATES AS FUNCTION OF SPECIFIC GROWTH RATES'
RP=GP*CXLO
RX=GX*CXLO
RS=GX*CXLO/YXS + GP*CPLO/YPS + AE*MS*CXLO
RO=GX*CXLO/YXO + GP*CPLO/YPO + AE*MO*CXLO
```

```
RC=GX*CXLO/YXC + GP*CPLO/YPC + AE*MC*CXLO
'CALCULATE THE ARRHENIUS EQUATION FACTOR FOR THE MAINTENANCE COEFF'
AE=(2.718**(4693.*(1/298. - 1/(TF+273.))))/2.
'CALCULATE THE SPECIFIC GROWTH RATES'
UMAX=(2.718**GROW(TF))/(10.**ABS(DEAD(PHL-7.,PHH-7.,PH-7.)))
GX=UMAX*(CSLO/(KM+CSLO))
GP=G1*GX + GO
'CALCULATE THE MASS TRANSFER RATES'
JO=RSW(COLO.GT.0.9*COLMAX,JOMAX,((YOI - YOO)*FGI*32.*PT)/(RG*(TF+273.)))
JC=RSW(CCLO.GT.0.9*CCLMAX,JCMAX,((YCO - YCI)*FGI*32.*PT)/(RG*(TF+273.)))
'CALCULATE THE GAS TO LIQUID MASS TRANSFER COEFFICIENTS'
KO=(2.*ZO/BD + 0.31*(NSO**(-2./3.))*(DPF**(1./3.)))*60.
KC=(2.*ZC/BD + 0.31*(NSC**(-2./3.))*(DPF**(1./3.)))*60.
'CALCULATE THE SCHMIDT NUMBER AND DIFFERENTIAL PRESSURE DRIVING FORCE'
NSO=(UF*0.01)/(DF*ZO)
NSC=(UF*0.01)/(DF*ZC)
DPF=((DF-DG)*UF*9.806)/(DF*DF)
'CALCULATE THE BUBBLE DIAMETER'
BD=4.15*(ST**0.6)/(((7.457E9*HP/VF)**0.4)*DF**0.2)*(VG/VF)**0.5 + 0.09
'CALCULATE THE BUBBLE SWARM VELOCITY'
BV=27.43*60. + 2.*FGI/AF + FA/AF
'CALCULATE THE BUBBLE RISE TIME'
BT=HF/BV
'CALCULATE THE AGITATOR PUMPING RATE'
FA=28312.*NQ*RPM*(RI/1.27/12.)**3
'CALCULATE THE GAS ,LIQUID, AND FERMENTOR VOLUMES'
VG=REALPL(0.5,FGI*BT,VGIC)
VL=INTEG(FSI*DS/DF,VFIC)
VF=VL + VG
HF=VF/AF
'CALCULATE THE INTERFACIAL AREA'
AI=(VG*6.)/(VF*BD)
'CALCULATE THE PHYSICAL PROPERTIES OF THE FERMENTATION BROTH'
UF=BOUND(0.01,10.,(UM*0.01)*(CXLO/KM))
DF=(1. + CXLO + CSLO)
CF=(1. - CSLO)*CC + CSLO*CL
'SIMULATE THE ENERGY BALANCE DYNAMICS FOR THE LIQUID PHASE'
TF=INTEG((QG + QA + QS - QJ - QV - QC)/(VF*DF*CF),TFIC)
'CALCULATE THE HEAT DUE TO CELL GROWTH'
QG=BOUND(0.,VF,(GX*CXLO*VF)*(HS - YXS*HX)/YXS)
'CALCULATE THE HEAT DUE TO AGITATION'
RN=10.754*RPM*((RI/1.27)**2)*DF/UF
HP=POWER(RN)*(RPM**3)*((RI/1.27)**5)*DF/1.523E13
QA=641616.*HP/60.
'CALCULATE THE SENSIBLE HEAT'
QS=FGI*DG*CG*(TG - TF) + FSI*DS*CS*(TS - TF) + FLO*DL*CL*(TL - TF)
'CALCULATE THE HEAT LOSS DUE TO EVAPORATION'
QV=HV*(CWGO - CWGI)*FGI
'CALCULATE THE HEAT LOSS TO THE COILS'
COIL(TCDT,TWDT,QC,QW,QF,TCO=TCI,FCI,DC,CC,VC,...
N,UA,TC,TW,TF,FCNORM,DW,CW,VW)
'CALCULATE THE COOLANT AND COIL WALL TEMPERATURES'
TC=INTVC(TCDT,TCIC)
TW=INTVC(TWDT,TWIC)
'CALCULATE THE FERMENTOR PH'
PHION(PH,CI=PP,WP,MW,HN,PK1,PK2,3)
PROCEDURAL (WP=CALO,CCLO,DF)
WP(1)=CALO/DF
WP(2)=CCLO/DF
END
'CALCULATE THE WEIGHT FRACTION OF CARBON DIOXIDE NOT IONIZED'
CCXO=CCLO/(10.**(PH-6.317) + 1.)
```

APPENDIX B

```
'SIMULATE THE INDUSTRIAL FEEDBACK CONTROLLER DYNAMICS'
CXSP=REMOTE(MINUTE)
CXCO=CNTRL(CXCOIC,CXC,CXM,CXSP)
DOCO=CNTRL(DOCOIC,DOC,DOM,DOSP)
PHCO=CNTRL(PHCOIC,PHC,PHM,PHSP)
TFCO=CNTRL(TFCOIC,TFC,TFM,TFSP)
'CALCULATE THE INFERRED CELL CONCENTRATION FROM A COMPONENT BALANCE'
OUR=(COGI - COGO)*FGI
CPR=BOUND(0.,1.,(CCGO - CCGI))*FGI*(10.**(PH-6.317) + 1.)
CXOM=INTEG((OUR - MO*CXOM)*YXO/VL,CXOMIC)
CXCM=INTEG((CPR - MC*CXCM)*YXC/VL,CXCMIC)
'SIMULATE THE MEASUREMENT TIME LAGS'
CXM=REALPL(CXMTC,CXOM*AAA + CXCM*(1. - AAA),CXMIC)
DOM=REALPL(DOMTC,DO,DOMIC)
PHM=REALPL(PHMTC,PH,PHMIC)
TFM=REALPL(TFMTC,TF,TFMIC)
'CALCULATE THE MANIPULATED AGITATOR SPEED'
RPM=(DOCO*(1.-CCC)+0.2)*RPMMAX
'CALCULATE THE MEASURED CONCENTRATIONS'
DO=COLO/DF*1.E6
'CALCULATE THE MANIPULATED FLOWS'
FSI=(CXCO*OOO + CXCOIC*(1. - OOO) + FFS*RAMP(180.))*FSIMAX
FGI=(DOCO*CCC + DOCOIC*(1. - CCC) + FFG*RAMP(180.))*FGIMAX
FAI=PHCO*FAIMAX
FCI=TFCO*FCIMAX
FLO=CXCO*FLOMAX*(1. - OOO)
'SET THE VARIABLE NAMES FOR THE OUTPUT PLOTS'
O2=COGO/DG
CO2=CCGO/DG
DILUTE=FLO
GAS=FGI
FEED=FSI
TEMP=TF
GROWTH=GX
CELL=CXLO
PROD=CPLO
FOOD=CSLO
END
'UPDATE THE MAXIMUM MASS TRANSFER RATES'
JOMAX=JO
JCMAX=JC
TERMT(MINUTE.GE.STOP)
END
TERMINAL
END
END
SET PRN=9
SET TITLE = 'FED BATCH  1500 LITER REG FERMENTOR TEST...
O2 UPTAKE RATE (OUR)-CO2 PROD RATE (CPR)'
PREPAR MINUTE,OUR,CPR,O2,CO2,DO,GAS,DILUTE,TEMP,GROWTH,FOOD,PROD,...
CELL,CXOM,CXCM,FEED,BD,BT,KC,JC,RC,KO,JO,RO,VG,YOE,YCE,PH,RPM,VL,...
QA,QG,QV
SET TCWPRN=72
SET NDBUG=1
OUTPUT MINUTE,OUR,CPR,O2,CO2,DO,GAS,DILUTE,TEMP,GROWTH,FOOD,PROD,...
CELL,CXOM,CXCM,FEED,BD,BT,KC,JC,RC,KO,JO,RO,VG,YOE,YCE,PH,RPM,VL,...
QA,QG,QV,'NCIOUT'=10
PROCED RUN
START
PLTDIN
SET CALPLT=.F.
SET STRPLT=.T.
```

```
SET TTLSPL=.T.
SET GRDSPL=.T.
SET XINSPL=10.
SET YINSPL=3.0
SET YTISPL=0.5
SET YCISPL=1.5
SET NPCSPL=10
PLOT 'XLO'=180.,'XHI'=STOP,OUR,CPR
SET TITLE = 'FED BATCH   1500 LITER REG FERMENTOR TEST...
O2 IN OFFGAS (O2)----CO2 IN OFFGAS (CO2)'
PLOT 'XLO'=180.,'XHI'=STOP,O2,CO2
SET TITLE = 'FED BATCH   1500 LITER REG FERMENTOR TEST...
DISSOLVED O2 (DO)---AGITATOR SPEED (RPM)'
PLOT 'XLO'=180.,'XHI'=STOP,DO,RPM
SET TITLE = 'FED BATCH   1500 LITER REG FERMENTOR TEST...
FERMENTOR PH (PH)-----TEMPERATURE (TEMP)'
PLOT 'XLO'=180.,'XHI'=STOP,PH,TEMP
SET TITLE = 'FED BATCH   1500 LITER REG FERMENTOR TEST...
GAS FLOW (GAS)------SUBSTRATE FEED (FEED)'
PLOT 'XLO'=180.,'XHI'=STOP,GAS,FEED
SET TITLE = 'FED BATCH   1500 LITER REG FERMENTOR TEST...
CELL CONC (CELL)----GROWTH RATE (GROWTH)'
PLOT 'XLO'=180.,'XHI'=STOP,CELL,GROWTH
SET TITLE = 'FED BATCH   1500 LITER REG FERMENTOR TEST...
DILUTION FLOW (DILUTE)--FEED CONC (FOOD)'
PLOT 'XLO'=180.,'XHI'=STOP,DILUTE,FOOD
DISPLY CXC,PHC,DOC,TFC
DISPLY QAIC,QGIC,QVIC
DISPLY RPIC,RXIC,RSIC,ROIC,RCIC
DISPLY TW
DISPLY TWIC
DISPLY TC
DISPLY TCIC
END
RUN
SET CMD=5
```